The Net-Works Guide to

WAP and WML

An introduction to the mobile Internet, and to creating your own WAP site

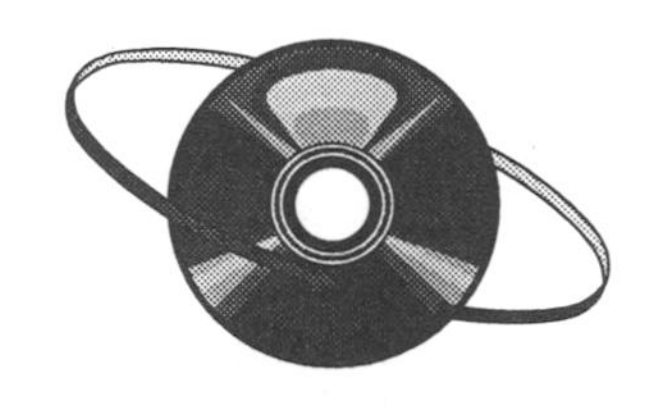

Pete Smith

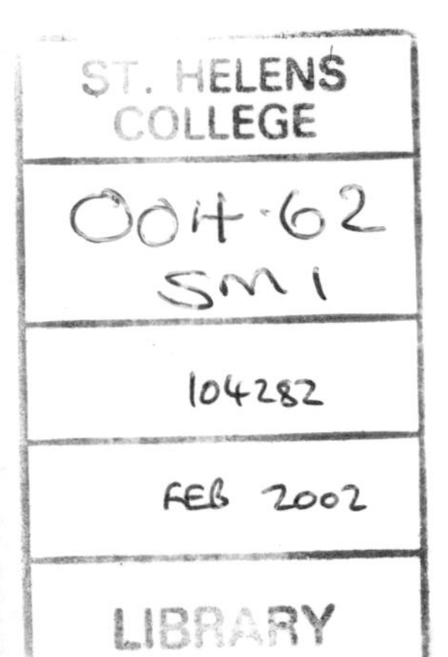

NET-WORKS

PO BOX 200
Harrogate
HG1 2YR
England

www.net-works.co.uk
Email: sales@net-works.co.uk
Fax: +44 (0) 1423 526035

Net.Works is an imprint of Take That Ltd

ISSN: 1-873668-91-0

10 9 8 7 6 5 4 3 2 1

Printed and bound in The United Kingdom.

Contents

***Introduction* 4**

***1. The Importance of WAP* 9**

***2. Open and Shut Portals* 15**

***3. Technical Aspects of the Protocol* 17**
a. Architecture
b. The WAP Stack
c. Functions of the Stack Elements

***4. Developing a WAP Site* 26**
a. The Toolkit
b. WML - Wireless Mark-up Language
c. WMLS - Wireless Mark-up Language Scripting
d. A Worked example

***5. Getting Heavy - Advanced WAP* 81**
a. Location Information
b. Pushing Yourself
c. User Agent Profiles

***6. Predicting the Future* 101**

***7. What is 3G?* 105**

Appendices

i. WML Attributes & Elements 110
ii. WMLS Standard Function Libraries 120
iii. XML - Extensible Mark-up Language 140
iv. MIME - Multi-purpose Internet Mail Extensions 145
v. Glossary & Acronyms 147

Introduction

This book is aimed at consumers and developers who want to explore the exciting possibilities of a world covered by access to the Internet via small hand-held devices.

These hand-held, or mobile devices include units such as mobile phones, PDAs (Personal Digital Assistants -which can be anything from glorified calculators to rather powerful handheld computers), even the smallest of the PC's.

The intention behind this book is to provide a guide with enough detail to for you to begin using wireless services via WAP, WML and its related components.

The *Net-Works Guide to WAP* is not a technical manual and will provide no more than a nodding glance (if that) at subjects such as the complexities of packet switching, the difficulties of routing via a cellular network or the physics of microwave data transfer. However, a basic knowledge of the Internet and website-creation is assumed.

If the IT industry is to be believed, more people will soon be accessing information via the mobile phone than the traditional desktop PC. Indeed sales of WAP phones are already matching and in some cases surpassing those of PCs. The arrival of Wireless Application Protocol (WAP) has gone some way to backing up this claim, by allowing the user to 'surf' the Net using a cellular phone.

What is WAP ?

The World is becoming connected by Mobile telephony. As the technology has matured developers have wished for the opportunity to go beyond voice communication and to harness the momentum of the Internet in order to provide an unimaginable range of services. WAP is the emerging standard which aims to try and deliver their wish.

Wireless Application Protocol is an open specification managed by the WAP Forum; meaning it is fully documented and, where no patents are established (tricky in this litigious world), anyone can implement the WAP standards. Hopefully this will tend to standardise the way mobile telephones access information and services.

Where has it come from?

Development of the Protocol is an ongoing effort, generally managed by the WAP Forum - a nominally democratic consortium backed, according to their statistics, by about 80% of the major players in this arena and many smaller but highly respected groups. The aim of The Forum is to provide a standardised environment for the provision of linked data services, such as the Internet, to the mobile community. The participants in the WAP Forum describe themselves as:

- Terminal and Infrastructure manufacturers
- Operators / Carriers / Service Providers
- Software houses
- Content providers, and
- Companies developing services and applications for mobile devices.

The board of directors contain luminaries from companies such as IBM, Nokia, Ericsson and Phone.com. The larger membership of the WAP Forum does seem to represent a broad range of interested parties, but it remains to be seen how the real authority within the organisation evolves.

The senior members such as Nokia and Ericsson do appear research-and-innovation based in their approach to bringing consumer products to market. Also the telecos such as BT and Vodaphone are gleefully adopting the technology albeit, in the opinion of some WAP gurus, a little cynically as it represents a late chance for them to recover from the fact that more and more voice services have been moving over to the Internet. Indeed Vodaphone have become so enthusiastic about the idea that a representative recently claimed that they "owned" WAP in response to a query as to why it was so hard to work out how to access services other than those provided by Vodaphone.

Other notable members or associate members of the WAP Forum include global telecomms giants such as Cable and Wireless, manufacturing outfits such as Motorola, Mitsubushi and Siemens along with Internet carriers such as UUNet. Software companies are represented by the likes of Phone.com and Macromedia, with smaller organisations such as Psion, the makers of handheld computers, and mapping company MapQuest all having a full say in the evolution of the WAP standards.

Where is it all going?

Future battlegrounds are already being marked out with dark rumours of litigation over technology rights emerging from time to time. Of course, any setup that has attracted the attention of such behemoths as Sun Microsystems, Oracle and (sharp intake of breath) Microsoft is bound to attract full and frank discussion. Indeed it could be that many are playing a cagey game, with more than one eye on stealing a march on the rest of the players.

Microsoft has an uneasy relationship with the whole WAP concept. Although it is a member of the WAP forum it is nonetheless pushing it's own proposal - known as Microsoft Mobile Explorer or MME. The organisation hopes to tie this in much more closely with it's own Windows-based vision of the future. With the software giant's track record of steamrollering standards this might be considered a real threat to the emergence of a consistent platform for mobile browsing of the web. However, you are probably safe in making choices now, because MML is expected to be fully WAP compatible.

Microsoft is betting on the bandwidth, battery life, screen and sound issues becoming less and less of an issue as it has with the land based Internet connections. Indeed the plans for the Third Generation (3G) of cellular communications suggest that hundreds of kilobits per second (rather than the current, rather-shaky tens of Kbits) are not too far in the future.

In theory, 3G technology will allow users to be connected to the Net each time their phone is turned on, making 'push' applications possible, which customise the information sent to each user automatically without them having to actually log-on.

Imagine you are dragged out to the shops on a Saturday afternoon when the big match is in progress. Your phone rings and tells you that your team have just scored. So you flip open your WAP device, which is using an 'always-on' connection, and you are able to watch a replay of the goal being scored. Alternatively, you may be near to a supermarket. This time your phone 'beeps' and relays a message from your fridge reminding you that you are low on milk and showing you some of the 'special offers' in the nearby shop.

Although it may sound that this technology some way off, but Cable & Wireless and Nokia of Finland have already completed the first phase of 3G tests. However, it seems likely that, for at least a couple of years, bandwidth will continue to be a major consideration and WAP will gain a larger share of the Internet connection market.

This book in a Nutshell

Chapter 1 kicks the book off with a look at **why WAP is important** to everyone who has an interest in the Internet or mobile phone business. Arguments will continue as to the longevity of WAP, but if you are missing out on this stage, you will probably miss out on what follows.

The second chapter looks at the current situation with regard to **service providers and independent portals**. You'll find an explanation of why certain parties are taking different stances, how you can circumvent restrictions that are put on your mobile surfing activities, and what is likely to happen in the future.

In Chapter 3, you'll find most of the **technical aspects of the protocol** explained. This isn't in a huge amount of detail, but it is designed to let you know what is going on behind the scenes. You'll find it fairly technical and is probably of more importance for those with a professional interest in WAP

Chapter 4 moves on to the **art of developing a WAP site**. You will be taken from the very beginnings and shown how to add different elements as you go along. The chapter ends with a worked example, section of which you can download from the book's web page.

The next chapter is where things get a bit heavy, with **advanced WAP functions**. Learn about vicinity information, push technology and user agent profiles. Don't forget your hard hat if you plunge into this.

To lighten the load a little, Chapter 6 takes a look at **what the future might hold** - it is quite exciting.

The book has a rather long tail with an **extended Appendix section**. At present, as you will have gathered, rather a lot of what there is to be said about WAP is rather technical. The appendices cover the driest of the technical aspects, but you'll find them a very useful reference point.

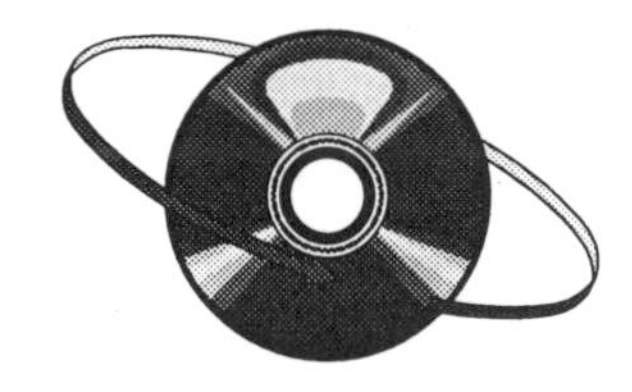

Chapter 1

The Importance of WAP

WAP is fast losing its staid and corporate image in the eyes of the public. Our ability to break away from the confines of the desktop PC, and access information 'on the hoof' is turning WAP into something of a watchword (or should that be acronym). The more we realise how much there is in it for us, the more popular it is becoming. The high street is simply groaning under the weight of 'promotional WAP-phones', Sunday papers are publishing articles on the best sites to visit.... the WAP generation is here.

The most important benefit of WAP is convenience. Actually the 'two' most important benefits are convenience and availability (but we'll stop there before this becomes a comedy sketch).

If you have any information on the Internet - whether personal web pages or a business site - there is a case for also providing it in WAP format. Most ISPs have configured their web servers to respond to WAP requests, meaning that any new WAP pages you create can live side by side with your conventional web pages on the same server. They even share the same upload mechanism. Even if you only have a simple homepage with details about your family, perhaps a few pictures of the kids opening their birthday presents so that Aunty Peggy in Australia can see their cherubic smiles, there is no reason why you can't post a few greetings or even your contact information in WAP format.

But beyond this, and well within your scope once you have read this book, why not create a WML deck (the official term for WAP pages) with your address book and phone numbers? With up to 15Mb of storage available on many ISP's you are able to hold considerably more information onsite than in any weedy little address book in a telephone's memory. You can even password protect it so that only you, or people you authorise, can access it. This comes into its own when you are out shopping, for example, and suddenly remember an important birthday. You need to post a birthday card but the last post is just about to be collected - and you don't have the address. Simply use your WAP phone to dial into your wapsite and look it up.

WAP is also of huge benefit to businesses - even the smallest company. You may own a shop selling specialist parts for washing machines, for example. Your customers can check to see if a part they need is in stock before calling around to collect it after their trip to the supermarket. Not only have you gained a customer by offering them instant, accurate information, you have saved time and manpower hunting around the store checking to see if the part is available.

This works both ways. If you are the one needing the spare part you save time phoning around to see if part XYZ is available, avoiding the frustration of having to wait around while the assistant searches the shop, only to come back and tell you they are out of stock!

Of course, if you have information that you feel might have an audience (other than yourself and Aunty Peggy) then the simple fact is that millions of people in the UK alone have WAP enabled mobile phones as you read this - you already have a market! This rapid growth of a technology that would ordinarily take time to filter down to the 'man in the street,' is largely due to the phone companies' persistence in pestering people to accept a free upgrade from the unit they already have, to a first generation WAP phone. Indeed, BT claim to have bought up the entire WAP phone manufacturing output of one of the worlds major phone producers. Worldwide there are hundreds of millions of WAP devices in eager hands, all searching for an excuse to be used.

At the moment very few people are using this technology to its full advantage - purely because most are not quite sure what to do with it. On your PC you can go straight to a search engine and look for whatever interests you at that time. As WAP uses gateways to access the infrastructure of the Internet, it is an ideal medium for searching and search engines. It is already apparent that most people who do use their WAP phone use it to find information about shops and services that are local to them, making it a valuable, if not essential, marketing tool for the small business and regional organisation.

One of the first things needed to get more WAP decks uploaded to the net (which will lead to more people using WAP) is for the hosting companies to not only make it easy to post content but for the many WML and WMLScript editors (software packages) to become as pervasive as the HTML editors are now. It won't be long before the main Web browsers will be able to render WML in some form, so the enthusiast working away in the spare bedroom or garage will be able to design WAP content rather more easily. Probably the main thing to bear in mind is that while the devices will rapidly become more powerful but that bandwidth will - at first - grow rather more slowly.

More than a Home Page

For medium sized endeavours, such as sites for small businesses, it is rather more important to ensure that the content is properly rendered on something more reliable than the current crop of rather idiosyncratic WAP browsers. It is worth pointing out, at this stage, that the WAP 'client' is not a physical device in itself, such as a mobile phone, but the combination of software and firmware inside such a device that uses WAP to perform its function. So,

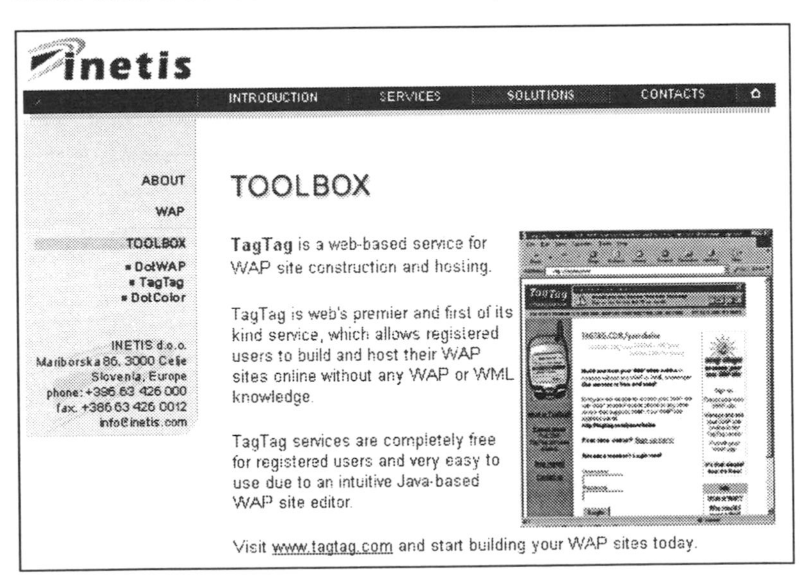

when a phone manufacturer adds the WAP acronym to its exhaustive list of functions in a bid to make you part with your money, it invariably means that a micro browser has been implemented. How that particular device then deals with the various control functions, or navigational issues, can be rather esoteric and as a result design issues are often compromised.

The most obvious commercial content is to provide contact details for a business such as a simple name, address and telephone/fax number list. However, it is simpler than you might imagine to add an email facility or perhaps an auto-dial out option. It is also relatively easy to add much more functionality as you'll discover later in this book.

You should bear in mind, however, that many of the current first generation phones have a level of RAM reminiscent of the early 1980s. There is no capacity in most of them for even medium sized data structures and most bloated modern programming techniques are unavailable. The KISS acronym (Keep It Simple, Stupid) is back in fashion. Bear this in mind and try to keep each card in the deck focused. Any diversity should be spread over separate, carefully managed decks - perhaps using some form of dynamic card generation.

Irrespective of whether you are a huge organisation or a dedicated content provider, you need to take the available hardware and the benefits of the medium into account, in addition to ensuring that there is value in the content. In the main this is likely to lead to concentration on providing access to services - especially convenience facilities such as buying CDs or ordering the delivery of a curry at 11 pm when you've had a drink. As the WAP model becomes more sophisticated, options to ensure efficient use of that most perishable of stock items - time - can be effectively implemented. Corporate Intranets could keep track of staff movements, and deliveries or collections lend themselves perfectly to being managed by WAP enabled devices.

The corporate giants who invest in WAP have shown that they are more than willing to throw money at any problem - speculative or otherwise. Indeed, the WAP Forum - a self-governing body of developers intent on ensuring that the 'new Internet' - which is effectively what WAP is - receives the undivided attention of the best brains in the business. Assuming that this has the desired effect of making the Internet accessible to the mobile user, as it undoubtedly will, then the whole balance of the Internet will shift towards wherever the new information centre of gravity sits. The current predominance of the desktop PC could well give way to equal, if not more, interest in services optimised for personal and mobile content.

For example, the music industry is struggling to come to terms with its semi-failure to capture, or rather control, the market on the Internet, albeit largely through its own greed and pricing models. If it fails, or assumes it is going to fail, on the 'normal' Web, the chances are it will choose not to pursue the option of 'on demand' music to the mobile user. Already, however, there are small start-up companies poised to step in where the giants fear to tread. If the luddites in Megadodo Record Publishing don't fully grasp the marketing opportunities available, then the artists - their bread and butter - will find they have no shortage of opportunities elsewhere.

Fulfilling Potential

While the elegant structure of the cards and deck analogy is quite straightforward, giving the advantage of helping the developer break up problems into nice small bites, the technology is more than capable of supporting quite complex and sophisticated applications. The simple set of rules which makes up the Wireless Mark-up Language are supported by a straightforward "Java-like" scripting language called, unsurprisingly, WMLScript, that allows quite powerful stuff to be developed for the WAP device. But this is not the powerful bit.

The real power comes from an almost slavish compatibility with the underlying structure of the Internet. The same techniques that Web developers have refined to develop the powerful commercial engines behind the massive growth of the World Wide Web, also work perfectly for WAP. Of course the display is a little cramped at the moment to say the least, but sophisticated banking applications are very active - and the financial giants use the same servers and underlying databases for their WAP decks as their Web banking.

Needless to say, current bandwidth and memory constraints ensure that the client side of the applications incorporates all the cunning and clever tricks that normal humans (i.e. not assembly language programmers) would prefer to pretend didn't exist for fear of deflating their egos. Developing WAP content is a classic case of simple components and rules combining to produce a complexity limited only by the imagination and intelligence of the minds involved.

WAP development should be as much for fun as profit. However, the lure of filthy lucre, while hardly the most noble of motives, will indisputably drive the WAP wave and offer several models for generating revenue. The most obvious is that generated by mobile phone calls. The margins will probably be kept high by the telcos - at least in the UK - while they claw back some of the astonishing sums of money spent purchasing 3G licences. Next come the premium rate services such as banking, where the customer pays - the 'clever' bit being that the customer is usually unaware of this as the cost is built into normal charges.

A more honest way of earning revenue is by generating trade geared towards mobile users - supermarkets can provide delivery services, hotels can be located and bookings taken. Subscription services, with or without advertising, can charge for up-to-the-minute news or sports results - replays of premiership goals moments after Manchester United have scored should be a winner in both Manchester and Bombay. In short there is almost an infinite number of ways for WAP to fulfil its potential, initially with entertainment and information, and secondly by encouraging whole new industries to emerge - simply by putting people in touch with their needs.

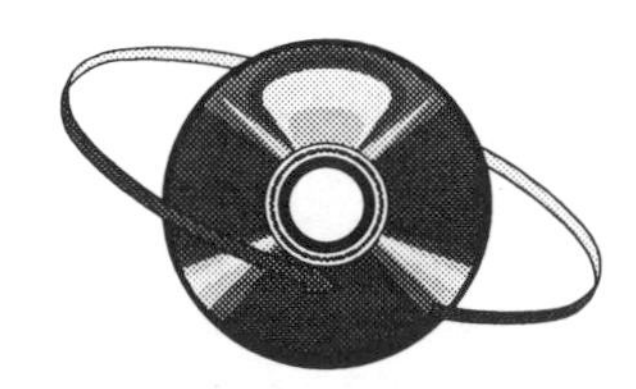

Chapter2

Open and Shut Portals

The infrastructure of the wireless section of the Internet is largely owned and controlled by a few lumbering giants, but still consolidation continuing. apace. With the likes of Vodafone swallowing the German giant Mannesmann whole without even pausing to chew, being a prime example. As this was also the first-ever hostile takeover of a German listed company, the signs for competition are not good.

If WAP had been structured in such a way that everyone would be forced to use a single allocated gateway to the 'wider' network, then the whole thing would have collapsed instantly - simply because of the prohibitive cost. Fortunately the WAP standard does not rely on bearer technology and it is perfectly feasible for anyone with the capital and knowledge to establish their own gateway to their own resources. This is just as well considering that the current generation of services are restrictive to say the least. Backed up by huge advertising campaigns, the main telecommunication companies would have us believe that the whole world of Internet access is easily available through a single point of contact - theirs.

But leaving aside the problem of finding WAP enabled content on the Internet (mainly because this is getting easier at an astounding rate) the 'official' portals are amazingly difficult to 'break out' of and you are easily led to believe that the services offered by your particular service provider are the whole thing. In fact the limiting of clients to 'back door' routes to the wider Internet shows how much money can be made from subscription services. It follows, therefore that independent content providers are critical in that they provide the essential competition needed against organisations that make Microsoft and Uncle Bill look like corporate benefactors.

As the cost of calls fall, and the various regulatory bodies force ever more open communication, it will progressively become more affordable for people with content to set up their own gateways and provide WAP windows into the Internet. Thus rather than being allowed to control the entire revenue stream, the mobile operators are being forced to allow other organisations to add value to the services in much the same way as the desktop-based services are being unbundled. At the moment the prohibitive cost of providing a gateway, coupled with the awkward set-up on WAP phones, make it easy for the big boys to hold onto most of the revenue.

The new WAP service provider, however, has many other ways of making money open to them. They could, perhaps, take a cut of the phone call cost. This is similar to the way in which Freeserve makes money from its tie-up with Energis. By simply employing advertising, subscription services, push services or even a combination of all three, money can be made from providing independent WAP services.

As one of the major uses of WAP will be to access corporate Intranets, independent gateways will be essential in providing secure commercial applications for procedures such as delivery tracking, stock management and sales force co-ordination. In this case, the developer could have a limited range of target devices - given the dogmatic mentality that thrives in some IT departments where the WAP client chosen is the one the MD heard about at the golf club.

On a positive note, the developer is guaranteed that any inadequacies will show themselves immediately and conclusively, forcing an almost instant change of provider. The obvious disadvantage, however, being that instead of the company keeping up a business advantage, which such new technology brings, the constant change will prove to be very expensive leaps forward, with previous developments being totally discarded.

This could result in a serious long-term problem in that WAP gateway providers may continue to restrict options in an attempt to maximise short-term revenue. Witness the stunningly high call charges which, until recently, cursed mobile phone users. Or the way BT priced ISDN so high that it was sourly referred to as "I Still Don't Need it". Indeed, they priced it so high for so long they effectively priced themselves out of the market - competitors and prospective customers are now bypassing it completely with newer ADSL technologies.

To cope with this possibility, the independent developer needs to stay abreast of as many new developments in the WAP sphere as possible.

Chapter 3

Technical Aspects of the Protocol

As mentioned in the Introduction, the aim of the WAP Forum is to provide a standardised environment for the provision of linked data services, such as the Internet, to the mobile community. There are several clear components to providing such a service, allowing it to interact end-to-end with the enormous number of resources currently available. The most obvious issues and components are...

a) **The Hyperlink or URL (Uniform Resource Locator) -**
As defined by the various Internet agencies and working groups. This is at the heart of the World Wide Web allowing as it does, the author of a document to provide links or 'jumps' between ideas in the same document or to a location in a totally different computer on the other side of the world.

b) **Typing of Content -**
On the web the content of a page is held in clearly defined datatypes such as text, pictures, moving images or sound, to allow browsers to render or view the page appropriately. If the various browsers on WAP devices are to use the same inform-ation then they must address the issues involved in maintaining the form of the page.
In many cases, at least initially, this will involve redesigning pages or simply avoiding tricky issues. As the technology becomes more mature you will see more faithful (and straightforward) use of the ever more complex datatypes.

c) **Markup Language -**
Internet browsers need to understand the intended appearance of the information - text needs to be laid out in a compre-

hensible manner, pictures, icons and functions need to placed and activated suitably. This is done with one version or another of a Standard General Markup Language (SGML). Currently the most common version of this is HTML (Hypertext Markup Language) with its emphasis on links, etc.

However there is a trend emerging towards the use of an extensible data-oriented language paradigm (not a phrase for the faint hearted - it simply means a special way of displaying different types of data) known as XML for short. This promises to allow far more sophistication on the one hand, and yet simplicity on the other, for web based applications.

d) Protocols -
The 'languages' of the underlying devices which determine who should get what response and routes the data to and from the various devices.

As advances in data communication push both the transmission capacity and the sophistication of the client devices to ever more powerful and useful levels, a need has emerged to optimise the communication between data source and the wandering device, whilst at the same time keeping the structure broadly sympathetic with the worthy ideals set out for the World Wide Web. Indeed the whole enterprise rests on the need to maximise the existing Internet standards wherever possible while providing a platform for as many difference type of service provider as possible. In this respect it is following the path to glory of the IP (Internet protocol) which had similar problems to overcome during the invention of ARPAnet - the forerunner of the modern Internet.

Needless to say this involves the marrying of many solutions to a range of technological issues. Many of these are beyond the scope of this book, but can be found in some detail in the 'WAP Architecture' document to be found on the website ***www.wapforum.org*** Only the basics follow.

WAP Architecture

The WAP model, or architecture, deliberately follows a broadly similar programming approach to that used in the World Wide Web. In the Web, the client device communicates either directly or indirectly with a server. It does this by passing a series of identification signals, followed by requests for data that the server interprets and responds to. These can be as simple as requests for a file to be transmitted or as complex as a sophisticated database query. The server then transmits the appropriate response to the request that is then rendered into a meaningful page (or some other type of response) on the browser.

Interaction can also be via some form of intermediary such as a proxy (often used to cache web information to avoid bombarding the servers) or a gateway, which may acts as a complex front for several servers or other data providing devices.

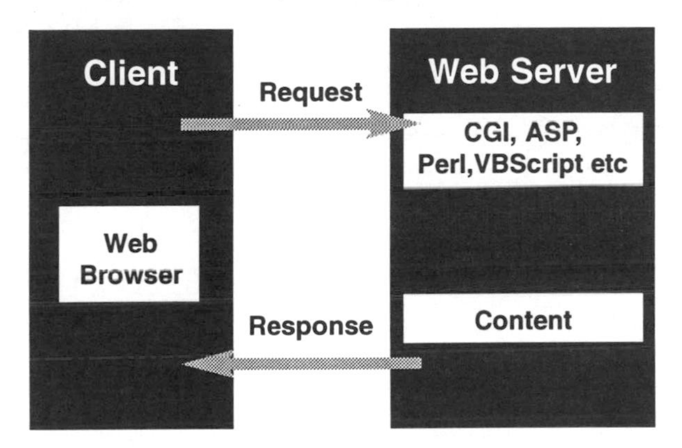

The most obvious feature of the WAP model is the need for more emphasis on this concept of a Gateway where messages are translated from the mobile device optimised WAP stack to the more general WWW protocols and vice versa. This allows applications to be developed using more familiar Web concepts and programming environments.

As well as responding to WAP content on WAP-enabled web servers, there is the possibility for communicating directly with a Wireless Telephony Application server (as opposed to a web server) that will also respond to the WAP client without translation.

Thus the WAP Model links the myriad steps of getting a request from the confined options on the WAP Client - or, more accurately, the software on the physical device - to the Internet. The various protocols involved help the messages survive the transmission all the way through to the different transmitters (as the client moves

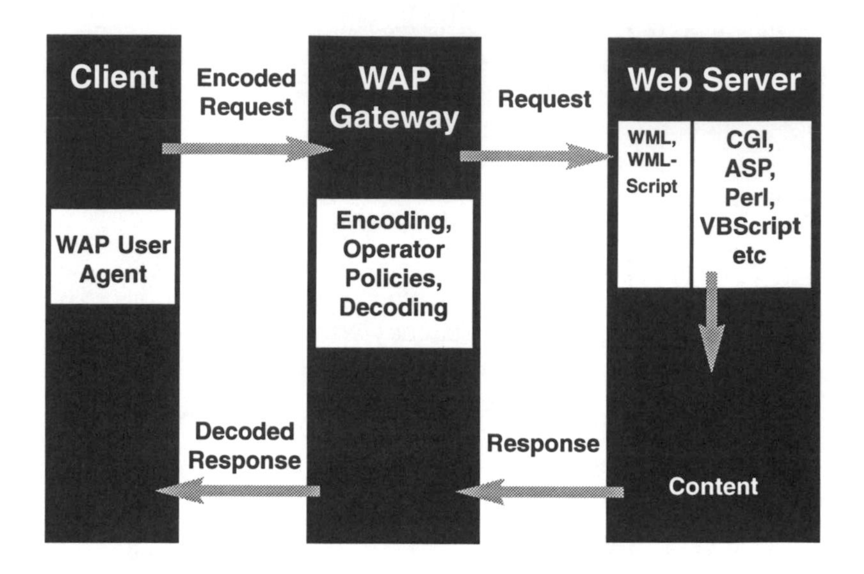

from cell to cell). The binary encoding (effectively a type of compression) keeps the messages as short as possible to conserve bandwidth and the precious processing capacity on the client. And the Gateway provides the processing that 'glues' the two worlds of Mobile telephony and the Internet. All of this may

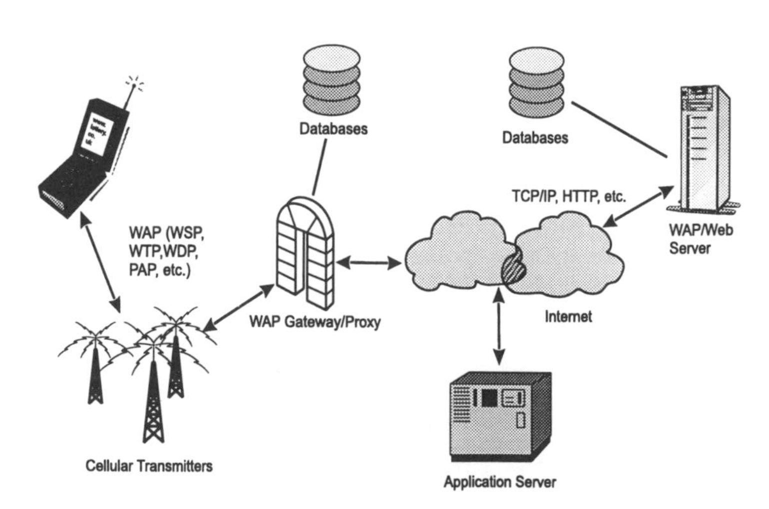

A WAP connection to the Internet from end-to-end

seem fragmented, but it is no less so than the Internet with its 'frame relay and routing and transmission issues (don't ask!)

The WAP Stack

The various protocols involved in the WAP connection sequence have a layered structure that is familiar to network managers and engineers all around the world, mapping as closely as it does to the OSI model of networking.

The OSI model is what baby Network managers are taught at nursery as a litany for the rules of networking. It Stands for "Open Systems Interconnect" and is based upon a standard proposed by the ISO (Organization Internationale de Standards). In practice the standard has never been fully implemented.

However it has been influential in the development of networks and distributed systems. For this reason it is important to understand the model. It is based upon seven layers, which are (from bottom to top)

- **The Physical Layer**, which defines the standards required for physical interconnections (the wire).

- **The Data Link Layer**, which defines the protocols for exchanging data frames over a "wire". This includes the important Medium Access Layer defining protocols for access to the shared wire of a LAN.

- **The Network Layer** is where routing take place.

- **The Transport Layer** provides for end to end connection between machines. Conceptually it is outside the network.

- **The Session Layer**, which provides for dialog control between processes.

- **The Presentation Layer** provides for translation of data structures between differing architectures.

- **The Application Layer** provides application level access to the network, file transfer, remote terminals etc.

Moving onto the WAP stack, you can see the similarities immediately:

OSI	TCP / IP	WAP		
Application (Layer7)	Application	Wireless Application Environment (WAE)		Other Services and Applications
Presentation (Layer6)		Wireless Session Protocol (WSP)		
Session (Layer 5)				
		Transaction Layer	Wireless Transaction Protocol (WTP)	
		Security Layer	Wireless Transport Layer Security (WTLS)	
Transport (Layer 4)	Transport	Datagrams (UDP/IP	Datagrams (WDP)	
Network (Layer 3)	Internet	Bearer Types: SMS, USSD, CSD, IS-136, CDMA, CDPD, PDC-P, Etc.		
Data Link (Layer 2)	Subnet			
Physical (Layer 1)				

Within this protocol stack the problems of sending and reconstructing data are broken down and solved one stage at a time. This approach allows programmers to utilise well-defined interfaces to access the features of the WAP stack.

The Functions of the WAP stack elements

Yes, things are about to get heavy...

Wireless Application Environment (WAE)

The main function of the WAE, which is based on Web technologies, is to provide an environment allowing the various operators and service providers to develop applications that can be rendered on the disparate collection of WAP enabled devices. It includes the following features:

- **Wireless Markup Language (WML) -**
  a lightweight markup language, similar to HTML, but optimised for use in hand-held mobile terminals

- **WMLScript -**
  a lightweight scripting language, similar to Java

- **Wireless Telephony Application (WTA, WTAI) -**
  telephony services and programming interfaces. Thus the WAP application can access the built-in functions of the unit such as telephony (remember that? - voice and stuff) and any other functions such as calendars, email and so on.

- **Content Formats -**
  a set of well-defined data formats, including images, phone book records and calendar information.

Wireless Session Protocol (WSP)

WSP allows the application layer to find a consistent interface to the connection-oriented service operating over the top of WTP or to the connectionless service above the datagram service WDP. At the moment the WSP layer consists of services geared towards browsing applications, exposing the following functionality:

- HTTP/1.1 functionality and semantics in a compact over-the-air encoding;

- Long-lived session state;

- Session suspend and resume with session migration;

- A common facility for reliable and unreliable data push; and

- Protocol feature negotiation.

The protocols in the WSP family are optimised for low-bandwidth bearer networks with relatively long latency.

WSP is designed to allow a WAP proxy to connect a WSP client to a standard HTTP server. Thus an ordinary Web server, when configured to support the correct MIME types, can publish to WAP devices.

Wireless Transaction Protocol (WTP)

The Wireless Transaction Protocol (WTP) runs on top of a datagram service providing a lightweight transaction-oriented protocol ideal for implementation in "thin" clients (mobile phones for instance).

WTP operates efficiently over secure or non-secure wireless datagram networks providing the following features:

- Three classes of transaction service:
  - Unreliable one-way requests,
  - Reliable one-way requests, and
  - Reliable two-way request-reply transactions.

- Optional user-to-user reliability - WTP user triggers the confirmation of each received message.

- Optional out-of-band data on acknowledgements - information outside the main communication stream.

- Asynchronous transactions.

Wireless Transport Layer Security (WTLS)

Complying with the Internet Protocol Standard Transport Layer Security (known until recently as SSL or Secure Sockets Layer) WTLS is designed to work well over the low bandwidth encountered by many WAP enabled devices. Its defined functions are:

- **Data integrity -**
  WTLS contains facilities to ensure data sent between the terminal and an application server is unchanged and uncorrupted.

- **Privacy -**
  WTLS contains facilities to ensure that data transmitted between the terminal and an application server is private and cannot be understood by any intermediate parties that may have intercepted the data stream.

- **Authentication -**
  WTLS contains facilities to establish the authenticity of the terminal and application server.

- **Denial-of-service protection -**
  WTLS contains facilities for detecting and rejecting data that is replayed or not successfully verified. WTLS makes many typical denial-of-service attacks which are common over the Internet, harder to accomplish and protects the upper protocol layers.

WTLS may also be used for secure communication between terminals, such as for the authentication of electronic business card exchanges. Applications are able to selectively enable or disable WTLS features depending on their security requirements and the characteristics of the underlying network (e.g., privacy may be disabled on networks already providing this service at a lower layer).

It should be noted that the WAP security model concentrates on the provision of a secure connection between the client and the server. However the emerging encryption standards and the recent relaxation of legislation on this topic by the US means that end-to-end security is likely to become commonplace as the applications are piggybacked onto the WAP protocols.

Wireless Datagram Protocol (WDP)

WDP acts as the Transport layer protocol in the WAP architecture and this layer operates directly over the data capable bearer services supported by the various network types. It is this that allows WAP systems to operate over any suitable type of Network. WDP allows the higher WAP protocols to find a consistent interface to the myriad types of bearer service. Thanks to this the Security, Session and Application layers are able to function independently of the underlying wireless network. This is accomplished by adapting the transport layer to specific features of the underlying bearer. By keeping the transport layer interface and the basic features consistent, global interoperability can be achieved using mediating gateways.

Chapter 4

Developing a WAP Site

The Toolkit

In order to develop and test your WAP service, you will need a WAP phone or an emulator with which to view your efforts. As with any new technology, you will probably find there will be a lot of work by trial and error. In other words, you'll need to look at what you've done on a frequent basis. If you have to dial-up a connection and wait for your WAP phone to display your site each time you change a line of code, you'll very quickly get frustrated. For this reason alone, it is suggested that you obtain an emulator - a piece of software that will allow you to view your WAP site via the web.

The examples in this book have been developed and tested with the Nokia Toolkit 1.3. which can be downloaded for free (an important consideration) after registration from the Nokia website (www.forum.nokia.com) in the developers area. You will also need to install the Java Virtual Machine 1.2.2 (Again a free download, available this time from Sun) to get the emulator to work.

The toolkit is both an emulator and a development environment. Although it tries to keep the feel of a windows application, its Java roots show through sometimes and screen updates can be a little frustrating at times.

In general however it is a well-specified environment and allows the aspiring developer to make a solid start in the world of WAP development.

The Emulator

In fact there are three emulation devices available in the Nokia toolkit, not one - two 'generic' phones, the 6150 and 6110 which are not 'real' devices, and an emulator for the 7110. This uses the code from the phone itself and to use the option you will need access to a gateway (Nokia also provide a downloadable time-limited gateway for NT based developers).

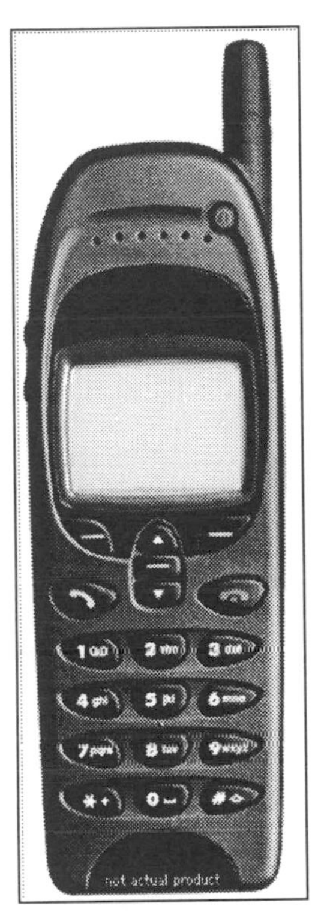

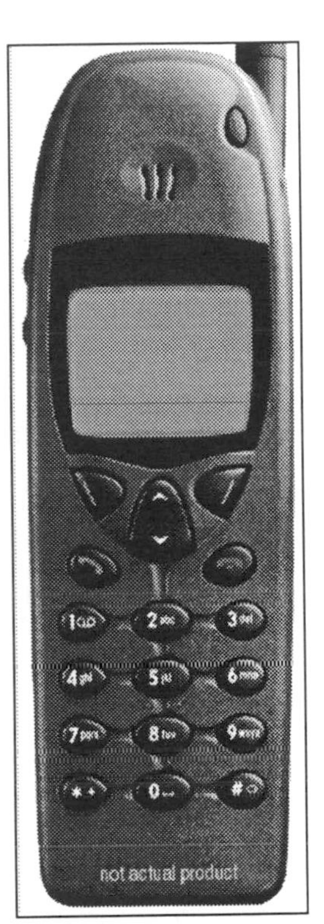

The functions are accessed through the buttons on the phone, either through the numeric pad for digits and letters, the * and other function keys or the special scrolling and selection keys, which are programmable by the WAP application on the unit.

As WAP devices evolve the default user interface is bound to advance, but at the moment on most WAP phones the soft keys choose options on the bottom line of the display. Scrolling through the options, or to view more of the current content, is achieved with the arrow keys.

Note, however, that but developers shouldn't assume that all users of their site will be presented with the same components to navigate with; other WAP interfaces might use a touch sensitive screen or variations on the theme of a scrolling wheel.

Equivalent to the mouse click is the select key, although the chosen option can usually be selected in some other way - the ergonomics of using such small devices suggest that this sort of short-cut will be common.

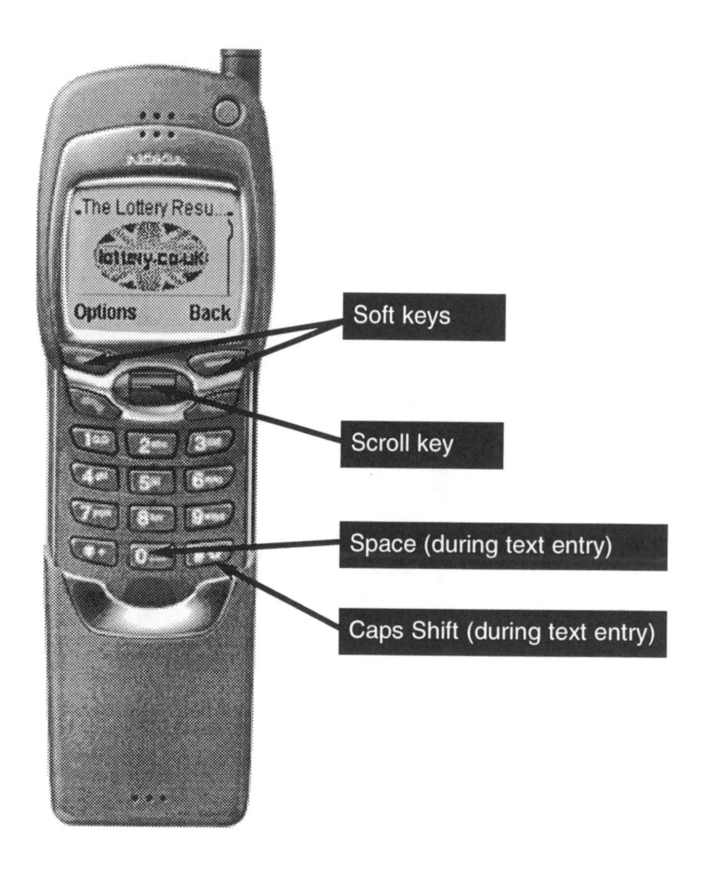

The Editor

The toolkit presents itself as a fairly standard editor. To start creating a deck you simply select 'file', 'new' and select either a new deck, script or wbmp image. The latter is very clumsy and should only be used as a last resort. However when selecting a new deck you are presented with an outline deck with the headers already nicely in place. Each new session is started in its own tab, allowing easy switching between the elements you might have open at any one time (i.e a deck and an associated WMLScript)

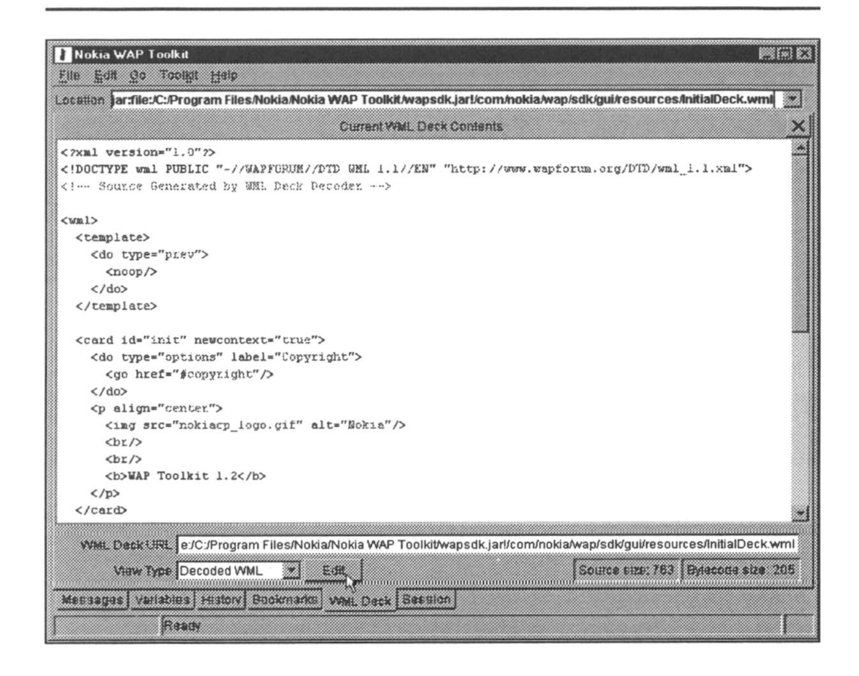

Similarly on opening a WMLScript file (i.e. filename.wmls) you are presented with a nicely laid out screen, again with the elements picked out in different colours.

As the text is entered the editor has a best guess as to the type of element and colours them accordingly. There's not much in the way of hand holding, though. As far as syntax is concerned, a missed or misplaced semi colon can still be totally confusing. That said, the help files are (despite being in Acrobat format) quite superb with just about everything you may want to know about WML and WML script being easy to find.

Having typed in your WML masterpiece and, perhaps, the attendant script and saved them, pressing the compile button causes the toolkit to parse the code. Depending on the encoding mode selected, the inevitable error messages will, once you get the hang of them, help you debug to some extent. Once you get a clean compile for your WML, selecting the 'show' button invokes the deck on the emulator.

At the same time the WML Deck tab shows the parsed source beautified (nicely indented etc.) allowing you to inspect the results of your compilation.

Other useful tabs include the **session tab** which maintains a list of all the links invoked so far in the session and the **variables tab**, which displays any variables currently in use and their current value. Should any errors crop up then these will be recorded in the **messages tab**.

Also useful is the **bookmarks tab** that allows you to preserve the URLs of you favourite WAP sites, including your own of course. Using this or the **'load location'** option under the **Go menu** item will invoke an HTTP connection to the remote file rather than simply loading the local file.

Although not perfect, this is at least a test of the way the deck should act under real conditions. It might also be an idea to use the compiled (*filename*.wmlc and *filename*.wmlsc) versions to ensure that your local encoder/decoder isn't doing anything odd. There haven't been many reported problems that aren't bugs (also called features once they reach production) in emulators, or in rare cases, the actual phones, but it doesn't hurt to be careful.

Other Ways to Develop Your WAP Site

While the Nokia toolkit is admirable in that it's quite functional and free, there are several other similar toolkits available from Phone.com, Ericsson and Motorola. More exciting is the entry of Macromedia into the development tools area. In association with Nokia they have introduced a graphical WAP site development kit that integrates into the flagship web development product 'Dreamweaver'

This uses the visual concept of adding a card then adding the various attributes by means of drag and drop, setting the values with simple menus. For example, to add a 'Do' event (Don't worry, these things are explained later in the book) the **Nokia_Do** object is dragged to the card or template, and the following form collects the attributes to determine how the component behaves.

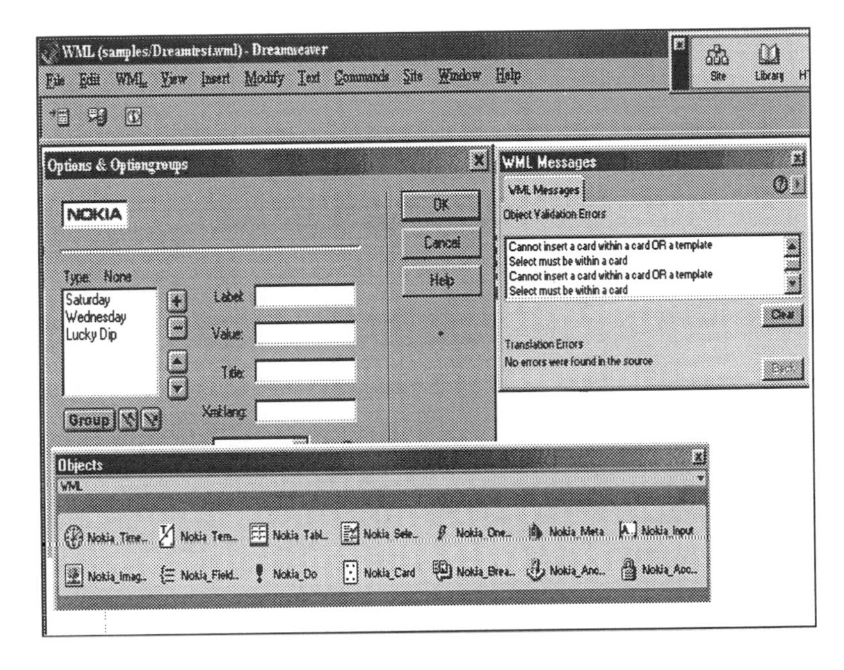

You also have the option of editing code directly with a rather nice editor

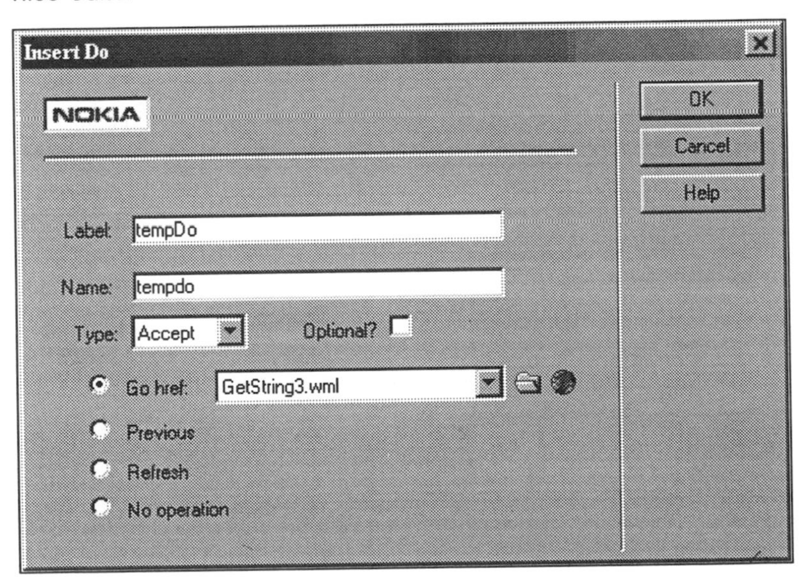

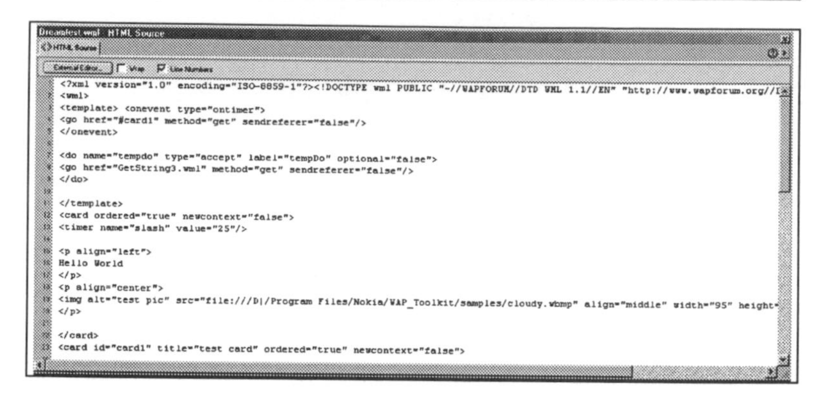

Although the Beta version displayed here lacks the colour information, so helpful in the original Nokia toolkit, it is quite probable that the final release of the Dreamweaver WAP add-in will become the de-facto standard for WAP development

Publishing your Site

Most web hosting ISPs have now configured their servers to handle WAP content, and simply putting the WML files in your web space for WAP devices to access will work to some extent. If you don't have convenient access to the Internet or a friendly local web server, or perhaps want to use dynamic content such as Microsoft's ASP technology, then Windows 98/2000 and NT 4 come with a Personal Web Server (PWS). This can - with a few minor tweaks - be persuaded to publish WML decks.

Although the documentation included with PWS is exceptional, covering most topics needed to get even a quite sophisticated Web site running, there is one feature you need to add before a WAP device can interpret PWS output. The issue is the type of document that the web server is sending you. This needs to carry a header identifying its MIME type - nothing to do with French acts locked inside imaginary rooms - the Multi-purpose Internet Mail Extensions are simply a standardised way of telling applications what sort of data they are dealing with (see the brief description in Appendix D). To convince PWS to do this you need to edit the registry (DON'T do this if you are not totally confident about what you are doing ***and*** your backups are safe. Messing up the registry can knadger your whole system to an unstartable state). Using Regedit or a similar tool you need to go to:

HKEY_LOCAL_MACHINES\SYSTEM\CurrentControlSet\Services\InetInfo\Parameters\MimeMap

Then right click in the right hand window, click "New - String Value" and type in (and repeat for the following):

text/vnd.wap.wml	modify enter in value data: wml
image/vnd.wap.wbmp	modify enter in value data: wbmp
application/vnd.wap.wmlc	modify enter in value data: wmlc
text/vnd.wap.wmlscript	modify enter in value data: wmls
application/vnd.wap.wmlscript	modify enter in value data: wmlsc

After the obligatory reboots that Windows always seems to need, your Personal Web Server should start publishing requests for WAP data types. If you are using the NT 4.0 version of PWS (or IIS) there is a more convenient access to this through the MMC, to extend the list of known MIME types.

Now, after a brief look at the Wireless Markup Language (WML) and WMLScripting language (WMLS), it's time to start developing a simple WAP site.

WML - Wireless Mark-up Language

The way to get a WAP device to show your wonderful and imaginative information source on its screen is via a 'Mark-up Language'. As the name implies, it originates from the world of printing where pages were 'marked up' in a standard manner known as Standard Generalised Mark-up Language. Thus a printer could - with a great deal of skill - define how various elements of the printed page were to be laid out by the computerised printing equipment. A headline size could be simply defined, a picture could be placed exactly where the author wanted it and text could be formatted to appear and flow in the style most appropriate to the article or whatever it was being printed.

The 'inventors' of the World Wide Web at CERN (a physics institute concerned with high energy physics - just where you'd expect the most significant technological revolution of our times to come from) liked the idea of a mark-up language, but decided that SGML was far too sophisticated and complex for the simple browsing technology available. Hence HTML was born, as an extended subset of SGML. As you know, it allows text to be linked to other parts of the same document or even documents on a totally different machine. The concept of hypertext, which had also been around for some years, finally had a natural home.

Then technology marched on with bandwidth and browsers becoming ever more sophisticated. And although HTML has undergone several revisions, it has become apparent that the huge number of data-types are beginning to choke the system and that a new, more flexible way of rendering data is needed.

The answer turns out to be the eXtensible Mark-up Language, or XML, with which the author can invent his or her own mark-up elements. The browser can then be instructed, via another consistent document, on how it should render these new elements.

At the same time, a whole new class of Web devices have come onto the scene with their own unique advantages and, from the point of view of displaying information, unique disadvantages. The body implementing the use of these new devices, the WAP Forum, was, from the outset, determined to avoid the mistakes of the various WWW proprietary bodies. So they opted to implement a suitably designed XML for the world of Wireless connection to the Internet.

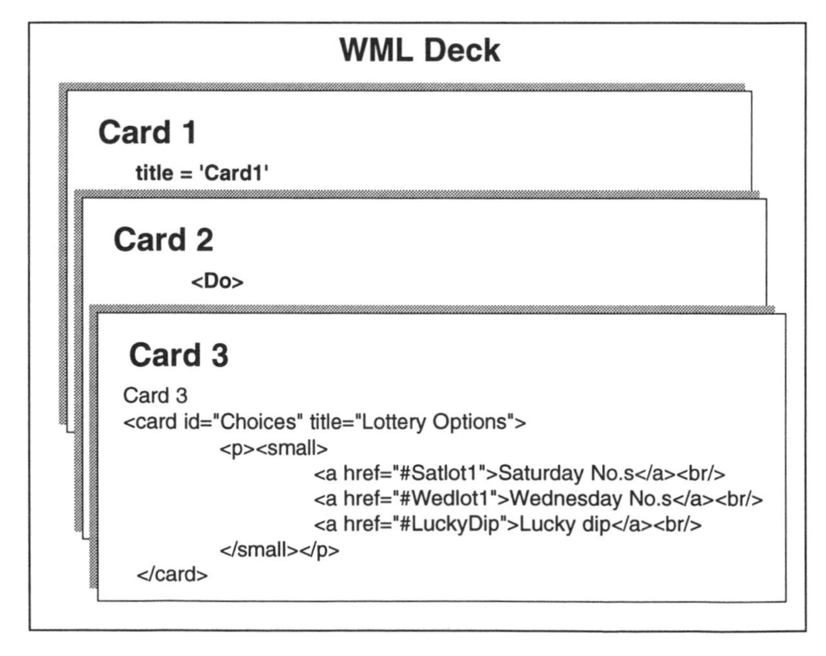

As a dialect of XML, the WAP Wireless Markup Language, or WML, consists of a list of instructions for the micro-browser to render information on the screen and to present options to the user for navigation of that information.

Organised using the metaphor of 'Cards' in a 'Deck', this (highly case sensitive) list consists of:

Elements

These specify the struc-ture and layout of the WML deck. They usually consist of a start tag, content and an end tag.

For example, the paragraph element `<p>` might be used to simply show a line of text

```
<p>The Lottery Results</p>
```

Attributes

Elements can have attributes that specify more information about that element. They are always contained in the start tag. It is worth noting that the attribute value is always wrapped in quotes to conform to the XML specification.

Here the align attribute is used to place the text in the centre of the display

```
<p align="center">The Lottery Results </p>
```

Comments

Although not displayed by the user agent, comments are intended to help the author maintain the WML Deck.

```
<!-- Template for the whole deck -->
```

Variables

The cards and decks can be modified at run time by variables that are identified by the prefix "$". For example, this will display the current value of the JustMaybe variable

```
<p>
$(JustMaybe)
</p>
```

The Structure of a WML Deck

WML operates at run time (i.e. when in use on the WAP device) in an environment referred to as a "Browser Context". This means all the parameters and status information are held at the client device. This Browser Context can be reset to a known state, if required by the author, to allow a "clean slate" for the WML deck or application.

As mentioned before, the deck structure follows the rules of a valid XML document. One of the first rules of XML is that there must be a valid XML header. That is to say, the very first part of the deck must indicate, in a formal way, that it is an XML document and it has the associated definition of how the document should be presented. If this 'preamble' is missing the deck will fail with some form of error. It typically contains something similar to the following statements:

```
<?xml version="1.1"?>
<!DOCTYPE wml PUBLIC "//WAPFORUM//DTD WML 1.2//EN
"http://www.wapforum.org/DTD/wml_1.2.xml>
```

The first statement declares the fact that what follows is XML and which version it complies with (XML is still very much a work in progress). And the second statement indicates where the associated Document Type Definition (DTD) can be found.

This is followed by the WML element, which defines the deck and encloses all the information

```
i.e.   <wml> all the stuff in the deck goes between
these tags </wml>
```

Here's a sample card that you'll be seeing more of in the example deck constructed later in this Chapter:

```
<wml>
<card id="intro" newcontext="true" title="The Lottery
Results">
     <p align="center">
        The Numbers are
        <br/>
    2 6 11 19 31 41
```

```
        <br/>
        <b>Good luck</b>
      </p>
   </card>
</wml>
```

The key elements in WML are the <do>content</do> and <go>content</go> tags. Each of these carries a range of attributes that allow the author to direct navigation in response to the user's input (more on those later)

The same example illustrates well the <card></card> element which is always encapsulated between the <wml> and </wml> elements. The 'card' element contains each displayed unit of content (each screen) and further encapsulates all the mark-up information for that content. Typically a card will carry some initialisation information to set up the display and any variables that are immediately needed. The example above gives the card an identification (so it can be found by navigation commands) of "intro", and a tile of "The Lottery Results" which is displayed at the top of the browser screen. It then displays the content within a single paragraph <p></p> couplet.

WML supports mechanisms for navigating the cards in a deck (and between decks) and event handling while a card is being displayed. The elements involved permit control over the way the content is displayed and how the user can move between options.

The simplest method available for navigating back through content is the History Model. Using the conventional model of a stack - think of it as a stack of cards - the pointer to the current card is 'pushed' on the top of the stack when the card is accessed and 'popped' to go back, thus providing a mechanism to easily skip back to the previous card at the press of a button. The tag which initiates this is <prev/>. At this point, note the trailing forward-slash. This is because the tag has no partner and therefore needs to 'close' itself.

The URLs (identifications written on the cards) in a history stack can be reset. That means the stack can be set to contain no addresses. A '<go>' event results in the current address being 'pushed' - where a new address is added as a new card is reached - before the jump to the new address is made. When a jump to a previous card is reached, i.e. the '<prev/>' element is encountered in the code, the stack is said to have been 'popped' and the top address is removed from the stack after it has been used to locate the destination of the jump. If you don't follow that first time, jump back to the previous paragraph, and repeat.

To address a jump, WML has adopted the HTML way of naming locations within a resource. A WML fragment anchor is specified by the document URL, followed by a hash mark (#), followed by a fragment identifier. WML uses fragment anchors to identify individual WML cards within a WML deck. If no fragment is specified, the URL names an entire deck, and the deck URL also identifies the first card in a deck:

```
<go href="#Choices"/>
```

Refers to the address of the card with the id 'Choices'

```
<go href="http://lottery.co.uk/index.wml"/>
```

Refers to the top card in the entire deck "index.wml" stored on the World Wide Web in the domain lottery.co.uk

Events in WML

There are various WML instructions that respond to events as diverse as entering a card or a timer expiring. There are four main elements in this category. They are ...

go

Effectively the <go> element results in a jump to a new destination, which is then displayed. As mentioned earlier this 'pushes' the address of the current URL on to the history stack, then navigates to the specified URL.

```
<do type="accept" label="Lottery options">
<go href="#Choices"/>
</do>
```

prev

Pops the history stack and navigates to the previous URL on the history stack.

```
<do type="prev" label="Previous">
<prev/>
</do>
```

refresh

Simply updates the display and variables associated with the current browser context.

noop

Specifies that nothing should happen (no operation) i.e. wait until something else interrupts the status quo.

Templates and Shadows

Currently the basic method for navigating a WAP device is through the function keys provided by the device. But whilst there is no compulsion in the WAP standards to offer any alternatives (they just say that a 'widget' must be provided to facilitate such choices) it is certain that sophisticated WAP devices will implement more convenient 'soft', keys - perhaps through buttons on the screen or even through voice control.

Templates allow an author to provide a default set of options for the whole deck.

```
<!-- Template for the whole deck (if nothing else is
specified) -->
    <template>
        <do type="prev" label="Previous">
                <prev/>
        </do>
        <do type="accept" label="Lottery options">
                <go href="#Choices"/>
        </do>
    </template>
```

However, it is also entirely possible for the author to specify alternative options for a specific card, which will override or "shadow" the settings outlined in the template with the local settings. The most common mechanism for the author to respond to the users choice from the options on the card they are on is via the "do" element, which takes the form

```
<do attribute list > Content </do>
```

Due to the wide range of devices that might be rendering the WML deck, the author cannot know how the do element will be displayed. It must, however, be assigned to some unique component which the user can access, such as a graphical button, menu item or function key. When the user activates the "do" element, the associated task is activated.

The anchor element specifies the head of a link. It defines the position in the page layout where a jump or other form of link is 'anchored'. There are two forms of the tag - one long and the other (unsurprisingly) short:

Long form

```
<anchor parameterlist >content</anchor>
```

Short form

```
<a parameterlist >content</a>
```

For example, the following will jump the user to the card with the id="Satlot1"

```
<a href="#Satlot1">Saturday No.s</a>
```

This element can be present in any text flow other than 'option' elements. The anchored link has a task associated with it, specifying the action to be followed when chosen.

So - remembering that nothing, no spaces, no carriage returns or whatever can come before the XML header - a very simple example of a WML deck might look like this.

```
<?xml version="1.0"?>
<!DOCTYPE wml PUBLIC "-//WAPFORUM//DTD WML 1.1//EN"
"http://www.wapforum.org/DTD/wml_1.1.xml">

<!--Start the Deck -->
<wml>

<!-- The First card -->
```

```
<!--Initialise the first Card-->
  <card id="Card1" title="www.lottery.co.uk">

<!-- Now program a soft key to navigate to the next
card -->
      <do type="accept" label="Next">
            <go href="#Card2"/>
      </do>

<!-- and display the content -->
    <p>
       Sat May 6
       6,11,19,38,38,40 Bonus Ball 33
    </p>
  </card>

<!--The second card -->

  <card id="Card2" title="www.lottery.co.uk">
      <do type="accept" label="back">
            <go href="#Card1"/>
      </do>
    <p>
       Sat Apr 29
       1,2,3,4,5,6 Bonus Ball 33
    </p>
  </card>

</wml>
```

Intrinsic Events

It is essential to be able to direct activity when the user makes a decision or something changes in the browser environment. Known as 'intrinsic' events these interruptions to the status quo allow the author to manage the user choices or to direct the status of the WAP device (the browser context) depending on how the user navigates through the deck.

The intrinsic events supported in WML are:

- **ontimer -**
  this can function in a card or template element and occurs when a timer expires.

- **onenterforward -**
  also at card or template level this event is triggered when a card is arrived at by the browser in the 'forward' direction.

- **onenterbackward -**
  triggers when the history stack is popped such as by a `<prev>` task.

- **onpick -**
  available in the option element, this event fires when an item is selected or deselected.

Here's an example of how it might look:

```
<onevent type="onenterforward">
   <go href="Lottery.wmls#LuckyDip('JustMaybe')"/>
</onevent>
```

Variables

WML allows the author wide flexibility in creating versatile decks through the use of variables. Variables can be used throughout a WML deck in the place of fixed data and are substituted at run time to allow data values to be set. A legal variable name begins with an underscore or a letter character, followed by zero or more letters, numbers and underscores. As with all WML issues the names are case sensitive.

The setvar element sets the specified variable in the current browser context to the required value. This command sets the variable called "Total" to a value of 1234:

```
<setvar name="Total" value="1234">
```

Input elements can be used to set the variable to any information entered by the user.

More Parts of the Deck

The <head> element of the deck allows the specification of information for the whole deck, such as meta-data and access control information.

The <access> element defines any restrictions as to which objects can access the deck. It does this by combing a domain and path attribute to build a 'mask'. Access is denied if a URL does not fit the mask. If no access tag is present, the default is that the deck is available to any card in any deck.

The <meta> element allows the declaration of generic information relating to the deck. It does not allocate any properties and the user agent is not required to interpret this data.

The <template> element as mentioned before, declares a template for the cards in a deck allowing the author to define a default set of event responses for every card.

The <card> element identifies the main subcomponent of the deck. Each card can contain a variety of content with the user agent rendering it depending on the agent capabilities and the type of content.

More about card intrinsic events

The <card> element can contain intrinsic events to allow a card to be prepared prior to display with conditions to be fulfilled. To illustrate this, the following snippet shows how the onenterforward attribute invokes the function 'LuckyDip' from the WMLScript file called'Lottery.wmls' immediately on entering the card.

```
<card id="LuckyDip" title="www.lottery.co.uk">
    <onevent type="onenterforward">
        <go href="Lottery.wmls#LuckyDip('JustMaybe')"/>
    </onevent>
    <p>
        <do type="accept" label="ReCalculate">
        <go href="Lottery.wmls#LuckyDip('JustMaybe')"/>
        </do>
        <do type="accept" label="Lottery options">
                <go href="#Choices"/>
        </do>
        <strong>Lucky Dip Suggests</strong>
```

```
        <br/>
                $(JustMaybe)

    </p>
  </card>
```

Another intrinsic event might be to display a splash screen for a few seconds using the timer attribute as follows.

```
<!-- First set up the card to jump to the #Choices
address in this deck -->

<card id="Logo" newcontext="true" ontimer="#Choices"
title="The Lottery Results">

<!-- countdown for 5 seconds in 1/10s of a second-->
<timer value="50"/>

<!-Display the picture and wait for the timer to expire
-->
<p align="center">
        <img src="lotlogo.wbmp" alt="Lottery"/>
        <br/>
        <br/>
        <b>The Lottery Results</b>
 </p>
 </card>
```

The card has several possible attributes with the option to set a title, describe the organisation of the content with the 'ordered' attribute, or initialise the browser context with 'newcontext'

Within a card there are elements that allow control over the presentation of the content (as far as is possible given the diversity of the possible target devices) and navigation between cards and decks.

The `<select>` element allows a user to choose an option from a list, each option being specified by an option element.

For example the following fragment would allow the user to choose which age group they belonged to. If the user (being honest) selected 'under sixteen' then the WAP application would set the variable 'Legal' to the value 'No' and might then inform the user that it was illegal to buy an on-line lottery ticket or order cigarettes or whatever services were on offer.

```
<wml>
 <card>
  <p>
Please enter your age group:
  <select name="Legal">
    <option value="No">Under Sixteen</option>
    <option value="Yes">Over Sixteen</option>
   </select>
  </p>
 </card>
</wml>
```

The choice of which type of lottery numbers were required is another obvious point at which an options list might be used i.e.

```
<p>
Select option
  <select>
    <option onpick="#Satlot1">Saturday No.s </option>
    <option onpick="#Wedlot1">Wednesday No.s </option>
    <option onpick="#LuckyDip">Lucky dip</option>
  </select>
</p>
```

However on many current devices the rendering of the options to select from can take several stages as the user chooses to edit selections then gets presented with the options to choose from. The maxim of "keep it simple" should be followed wherever possible (along with "don't get twisted and bitter" when it doesn't work)!

It should also be noted that the select and option elements can be combined in far more complex combinations by using the attributes available to the select element. For example, you could create multiple choice and default values.

The **<input>** element is used to specify an entry object to take input from the user. The input may be limited by the extensive list of options for the 'format' attribute and the system either inhibits the user from entering invalid characters (say lower case letters) or simply informs the user when input deviates from the expected.

This example assigns the input, which has been formatted to permit any number of upper or lower case letters only (no numerics), then allows the user to link to a card that will display a personal message.

```
<card id="InputName" title="what's your name?">
<p>
Please enter your name:
<input name="StaffName" type="text" format="*M" />
<br/>
<a href="#nextcard">Go to your Personal
message</a><br/>
</p>
</card>
```

The **<fieldset>** element is provided to assist in grouping associated inputs for display allowing the author to try and keep several choices and inputs on a single card. For example:

```
<card id="card1">
<p>
<do type="accept">
<go href="#card2"/>
</do>
<fieldset title="Name">
First name: <input type="text" name="fname"
maxlength="32"/>

<br/>Last name: <input type="text" name="lname"
maxlength="32"/>
</fieldset>

<fieldset title="Info">
<select name="sex">
<option value="Girl">Female</option>
```

```
<option value="Boy">Male</option>
</select>
<br/>
Age: <input type="text" name="age" format="*N"/>
</fieldset>

</p>

</card>

<card id="card2">
<p><small>
Hello $(fname) $(lname)
you are a $(sex) aged $(age)
</small></p>
</card>
```

Talk to Me

In many applications the server will want a response from the client in the form of user choices or entered information such as names, credit card details etc. This can be transmitted within a **postfield element** as part of the communication with a URL. For example, the following 'go' element would cause an HTTP GET request to the URL "//www.lottery.co.uk/purchase.asp?ticketNo=5" indicating the user had perhaps chosen to purchase 5 tickets through the on-line purchasing option.

```
<go href="//www.lottery.co.uk/purchase.asp">
<postfield name="ticketNo" value="5"/>
</go>
```

The relative merits of the various techniques for sending information via the Internet are a shade beyond this book. However it boils down to the 'Get' system encapsulating the data in the header which is easy to do but tricky to secure, and the 'Post' method which is harder to get right but puts the data in the body of a response, allowing more security mechanisms to protect the data.

Text

The rendering of text from WML is highly dependent on the client device; with many early devices not supporting all the text emphasis elements or combining them so, for instance, strong and bold have the same effect. The emphasis elements available in WML are:

- **em** - render with emphasis
- **strong** - render with strong emphasis
- **i** - render with an italic font
- **b** - render with bold font
- **u** - render with underline
- **big** - render with a large font
- **small** - render with small font

The paragraph **<p> element** and its attributes define the line wrapping and alignment for a paragraph. The default is that the text will break at any inter-word space and be left aligned.

The **
 element** forces a carriage return and a new line i.e. the next output will go to the beginning of the next line. This element is unusual as it doesn't occur as a <tag></tag> pair but simply as
 (closing itself in the same manner as <prev/>)

Setting the Table

The **<table> element** provides the author with a mechanism to create the layout of columns of aligned text or images. The **<table>** attributes along with the elements **<tr>** and **<td>** provide a fairly complete set of options for laying out simple tables and populating table them. Nesting tables is not allowed (outside cheap living room furniture).

The following listing should present a three-column table with various empty cells. Note that in the third row the browser is obliged to make its best stab at rendering the cell with a line feed in it. Here it should be realised that the required best effort is **not** the same as definitely coping with it.

```
<wml>
<card>
    <do type="accept">
```

```
        <go href="#card2">
                <setvar name="one" value="One"/>
                <setvar name="two" value="Two"/>
                <setvar name="three" value="Three"/>
                <setvar name="four" value="Four"/>
                <setvar name="five" value="Five"/>
                <setvar name="six" value="Six"/>
                <setvar name="extra" value="more"/>
        </go>
    </do>
</card>

<card id="card2">
<p>
<table columns="3" align="LL">
<tr><td>$(one)</td><td>$(two)</td><td>$(extra)</td></<
/tr>
<tr><td>$(three)</td></tr>
<tr><td>$(four)</td><td>$(five)<br/>$(six)</td><td>$(e
xtra)</td></tr>
</table>
</p>
</card>
</wml>
```

The **<pre> element** defines enclosed text as preformatted, requiring the browser to attempt to retain white space, use a fixed pitch font and disable any word-wrap.

As indicated earlier, WAP Provides for images to be displayed on the screen despite the somewhat pokey nature of some of the displays. Images are available in WML using the WBMP format, rendered with the **<img> element** and invoked from a URL. The attendant attributes of the <img> element prepare the user agent with information about its size and position while the image loads (as far as is possible). Some **special characters** have a particular meaning in WML, and cannot therefore be used directly in a WAP deck or card to represent themselves. To get around this, the following list shows what can be substituted in the WML code to result in, for example, the 'Greater than' symbol.

Entity	Notation	Explanation
quot	"	quotation mark
amp	& #38	ampersand
apos	'	apostrophe
lt	<	less than
gt	>	greater than
nbsp		non-breaking space
shy		soft hyphen (discretionary hyphen)

For example the following snippet

```
<p>
Obviously 1012 &#62; 611
</p>
```

should render

Obviously 1012 > 611

on the screen of the WAP Client

And that - in a nutshell is the basics of the Wireless Markup Language. It may look limited, but, when combined with the WMLScript scripting language, it can be used to develop useful applications. Furthermore, when used with server-side activity to generate dynamic WML for harnessing databases, some genuinely powerful systems can be built to deliver services with the potential to change the way we live.

WMLS - Scripting for WAP

Wonderful as WML is, it is inherently static, and it can only draw the page based on the instructions listed in the WAP pages. It doesn't directly provide facilities to act on the content it marks up and this could rather limit the value of the WAP model. The solution is, of course, WMLScript, which effectively extends WML to allow the author to build appropriate intelligence into the client end of the WAP service. Using WMLScript, the client device can be programmed to parse user input, utilise the capability of the client device and to configure the device. Most importantly it allows for dialog with the user in order to collate input and navigate the content without continual reference to the origin server - thus conserving bandwidth.

WMLScript is not based on the Beginners All-purpose Symbolic Instruction Code (BASIC) but it is an extended subset of JavaScript. Based on the snappily named ECMAScript a standardised form of the scripting language developed by Netscape and superficially similar to the Sun Microsystems Java Language, it has been modified to optimise bandwidth issues with a standard tokenised form among the refinements. It is defined as a procedural language with a comprehensive range of built-in and standard library functions. This means that programming is usually done in nice easy chunks and then stitched together, rather than in a single, long, straggling, incomprehensible mass (now, now, please don't be nasty about my writing style)! The similarity in syntax and structure to JavaScript means that many developers will find the task of learning WMLScript is a fairly trivial exercise.

The Lexical Structure (What does that mean?)

This is the technical set of rules that specify how a WMLScript is written. For example, the whole of WML and WMLScript is case sensitive so a variable referred to as 'DesignList' is entirely different to 'designlist'.

As with most programming languages, WMLScript is presented as a list of instructions with various options to allow the flow through the list, to branch out, or to jump to other parts of the list. Each complete instruction is known as a '**statement**', which is usually composed of one or more keywords and some form of evaluation or expression. Unless the statement explicitly states that a branch of some form should occur, then the next instruction in the list is processed until either the list is completed or an error occurs.

In keeping to its roots in Java, statements in WMLScript are terminated with a semi-colon. For example, the statement:

```
var Fourth=Lang.random(49)+1;
```

assigns a random number between 1 and 49 to the variable 'Fourth'. Hours of endless fun (frustration) at scripts that don't run, are quite likely to have been caused by a single obscure, missing or misplaced semi-colon.

WMLScript is defined as having three elements:

- **Variables** which hold values;
- **Functions** which combine values in various complex ways; and
- **Pragmas** which mysteriously prepare the compilation unit to understand various settings such as substituting an external script file with a local name.

These are referenced by "identifiers" which can begin with a letter or underscore but not a number.

Variables and stuff

Whenever programmers get together (and sometimes even singly) you will hear muttering about "data typing".

Put simply, "data typing" means how strict the programming language is about keeping data separate. There are good arguments in favour of 'strong typing' for languages used in heavy duty applications such as accounting or stock control where the slightest misunderstanding between the various components could be disastrous. Thus 'strong typing' means, for example, that a character such as '4' is totally distinct from the number represented as 4, and various hoops have to be jumped through to combine them.

In the interests of usability, however, WMLScript is described as a 'weak typing' language. This means that there are little in the way of rules about rigid data types. Indeed it is easy to use most variables to hold text (known as strings) when it suits you or numbers when appropriate.

There are five basic data types:

- **Boolean** (just 'yes' or 'no', True' or 'False', 'on' or 'off', 1 or 0 - however you like to think of it);
- **Integer** (no decimal points);
- **Floating** point (with decimals;

- **String** (text); and
- **Invalid** which is available to distinguish an invalid data type.

Any variable must be declared before it can be used, and its value persists (its scope) for the duration of the function it is declared in.

Convert or Die!

It's not quite as serious as that, but there is often a need to combine different data types such as building a string from a combination of characters and the values held in variables or constants. So although the language is weakly typed there is a need to convert values between the five main data types (or four if you discount the invalid type), which are handled internally.

If an operator needs a particular data type the system attempts to coerce the value into the correct data type automatically. Thus an integer will be converted to a string if a string operation is attempted and a Boolean 'true' is converted to a string with the value 'true'.

Operators and expressions

All the common operators are available in WMLScript along with a few less so. Of the more frequently encountered, the following list explains what they do.

+	add (numbers)/concatenation (strings)
-	subtract
*	multiply
/	divide
`div`	integer division
=	assign (NB: this does not mean 'equal to' as in some programming languages, this symbol makes the left hand variable equal to the result of the expression on the right.)

for example

```
age=21+18 results in age being set to 39
```

and

```
RealAge="Thirty"+"Nine"
```

sets the contents of the variable RealAge to "Thirty Nine"

Combined Operators

To save time and effort (and probably bandwidth) many operators occur in a combined form:

- `==` tests if two items are equal
- `+=` add (numbers)/concatenate (strings) and assign
- `-=` subtract and assign
- `*=` multiply and assign
- `/=` divide and assign
- `--` pre-or-post decrement
- `++` pre-or-post increment
- `div=` divide (integer division) and assign
- `%=` remainder (the sign of the result equals the sign of the dividend) and assign

Comparison Operators

- `<` less than
- `<=` less than or equal
- `==` equal (not assign)
- `>=` greater or equal
- `>` greater than
- `!=` inequality

Logical Operators

- `&&` logical AND
- `||` logical OR
- `!` logical NOT

If you really must know, the logical operators work as follows...

The logical **AND** operator evaluates the first operand and tests the result. If the result is false, the result of the operation is false and the second operand is not evaluated. If the first operand evaluates to true, the result of the operation is the result of the evaluation of the second operand. If the first operand evaluates to invalid, the second operand is not evaluated and the result of the operation is invalid.

Similarly, the logical **OR** evaluates the first operand and tests the result. If the result is true, the result of the operation is true and the second operand is not evaluated. If the first operand evaluates to false, the result of the operation is the result of the evaluation of the second operand. If the first operand evaluates to invalid, the second operand is not evaluated and the result of the operation is invalid.

...got that?

Bit Operator Operation

<<= bitwise left shift and assign
>>= bitwise right shift with sign and assign
>>>= bitwise right shift zero fill and assign
&= bitwise AND and assign
^= bitwise XOR and assign
|= bitwise OR and assign
~ bitwise NOT

There are various other slightly exotic operators available such as the **type of** operator that returns the data type of any given expression. It is also worth noting that WMLScript doesn't support arrays directly. There is, however, array-like functionality in the String library functions that allow the author to perform similar indexing and slicing actions using strings.

Some operators require a reference to a variable as the left operand. Thus another data type variable is used to specify that a **variable** name is needed.

WMLScript also supports most of the expressions supported by other programming languages, ranging from straightforward constants and variable names which simply evaluate to the value of the constant or the variable to more complex expressions defined by a combination of simple expressions, operators and function calls.

Functions

The basic unit of WMLScript, the function is defined within the WMLScript compilation unit (the file holding the script - remember that the code is compiled into a binary form en-route to the WAP device) and is used to perform a given set of instructions and to return a value (although if not specified this defaults to "").

The function is invoked or called either from a WML statement which provides the path of the compilation unit and any required parameters. For example

```
<go href="Lottery.wmls#LuckyDip('JustMaybe')"/>
```

Calls the function 'LuckyDip' in the local file 'Lottery.wmls' and passes the parameter in the variable 'JustMaybe'

Within a compilation unit, at least one function must be declared as accessible externally using the extern keyword, The parameters of the function are passed from the calling statement to the code by substituting the values provided into the parameter list of the function - a process known as 'passing by value' and the invoking call must provide exactly the corresponding number of parameters. These are then treated as "pre-declared" local variables. In addition to the inherent functions and any created by the author, there is a collection of function libraries available. These must be explicitly named when being called.

```
extern function LuckyDip(JustMaybe) {
    var Nums="xxxxxxxxx";
    var First=Lang.random(49)+1;
    var Second=Lang.random(49)+1;
    var Third=Lang.random(49)+1;
    var Fourth=Lang.random(49)+1;
    var Fifth=Lang.random(49)+1;
    var Sixth=Lang.random(49)+1;
    }
```

Statements

The 'substance' of a function is the sequence of the statements it contains. A statement consists of a combination of expressions and keywords linked to derive (hopefully) a result. Briefly (there is much more detail in the example) - in WMLScript the possible statements are:

var

Declares a variable for use, either by initialising it or assigning it a value.

For example:

```
var Fifth=0
```

expression statement

Assigns a value to a variable, evaluates mathematical formulae, makes function calls and so on.

eg.

```
Fifth=Lang.random(49)+1;
```

if...else

Used to branch conditionally to one or two statements.

eg.

```
        if (First > Second){
                SwapFlag=1 ;// at least one more loop
                SwapVal=First;
                First=Second;
                Second=SwapVal;
        };
```

while

Creates a statement that loops until the expression is no longer true.

eg.

```
    while (Second == First) {
                Second=Lang.random(49)+1;
                };
```

for

Creates a loop but holds the looping parameters in the initialising list.

eg.

```
for ( n=1; n<6; n++){ //all sorted after five passes
        ...repeated code in 'block' statements...
        };      //End for loop
```

break

Used to break out of a while or for loop.

eg.

```
if (!SwapFlag){
```

```
        break ; //if SwapFlag has reached zero stop
        };
```

continue

Breaks out of a block of statements in a "while" or "for" loop but does not break out of the loop. In a "while loop", it jumps back to the condition. In a "for loop", it jumps to the update expression. eg.

```
if (!SwapFlag){
                continue ; //next time round the loop
        };
```

block statement

A list of statements held in curly brackets, it is used wherever a statement is needed i.e. a while statement.

return

Defines the value returned to the calling function

empty statement

Used wherever a statement is required but no action is needed.

Although the basic number of keywords in WMLScript is quite limited it is expanded dramatically by the use of 'Libraries', each of which contains a collection of very useful functions. There are several standard libraries and in due course there are bound to be various proprietary libraries turning up as the manufacturers vie to provide the most powerful WAP device.

Libraries

The WAP definition of WMLScript defines the following libraries

Lang

A set of functions closely tied to the core of WMLScript:

abs	isInt	exit
min	isFloat	abort
max	maxInt	random
parseInt	minInt	seed
parseFloat	float	characterSet

Float

This library contains a set of helpful floating point functions:

int	pow	maxFloat
floor	round	minFloat
ceil	sqrt	

String

Contains various string functions which also allow a certain degree of array like activity:

length	elements	trim
isEmpty	elementAt	compare
charAt	removeAt	toString
subString	replaceAt	format
find	insertAt	
replace	squeeze	

URL

Contains a set of functions to handle relative or absolute URLs and read content from a URL:

isValid	getParameters	resolve
getScheme	getQuery	escapeString
getHost	getFragment	unescapeString
getPort	getBase	loadString
getPath	getReferer	

WMLBrowser

Provides functions allowing WMLScript to access current WML context:
getVar

setVar	prev	getCurrentCard
go	newContext	refresh

Dialogs

Contains a set of common user interface functions:

prompt	confirm	alert

For a full explanation of functions and their syntax see the standard library functions in Appendix B.

The purpose of this section has simply been to illustrate what functionality is available to the WMLScript. There are many advanced aspects not covered here, but the best way to get an understanding of how WML and WMLScript can be used together is to develop a small WAP application.

An Example Deck - Lets build a house of cards

To save you typing in this example, you will find all the code on the web page: www.net-works.co.uk/wapsample1.htm

To try to illustrate the various points of the art of building a WML deck, the aim is to build a simple WAP application to present the lottery numbers for recent Saturdays and Wednesdays and to navigate between them. To make it a little more interesting there will be an option to get a selection of randomly generated numbers to help the user win the jackpot.

The XML Header

To start the deck we need to conform with the niceties of producing a well formed XML document. Therefore the header needs to declare the nature of the document and the Document Type Definition.

```
<?xml version="1.0"?>
<!DOCTYPE wml PUBLIC "-//WAPFORUM//DTD WML 1.1//EN"
"http://www.wapforum.org/DTD/wml_1.1.xml">
```

Then on to create the WML deck proper.

The Deck Preamble

The initial task is to enclose the deck in the `<wml> </wml>` element.

```
<wml>
<!-All the rest of the deck goes here -->
</wml>
```

... then to provide a template for the deck. In this case you program the soft keys with default behaviour to allow both a consistent interface and reduce the need for explicitly defining navigation activity for each card. For example...

```
<!-- Template for the whole deck (if nothing else is
specified) -->
    <template>
       <do type="prev" label="Previous">
              <prev/>
       </do>
       <do type="accept" label="Lottery options">
              <go href="#Choices"/>
       </do>
    </template>
```

The above listing defines a template for the deck that defaults to allowing a choice between going back to the previous card (the `<prev/>` element) or jumping to the main menu card for the application (defined by `<go href="#Choices"/>`).

The First Card

Now for the first card in the deck! It is becoming something of a convention to present a logo for a few seconds. Fortunately this is quite an easy operation with images being stored in wbmp format and the existence of a card attribute 'ontimer'.

The timer is invoked by setting a countdown value which causes flow to pass to the card defined in the 'ontimer' attribute on expiry. In this case, the card is given the id of "logo" and told to display the lotlogo.wbmp graphic file for 5 seconds.

```
<!-- First card is a logo -->
  <card id="Logo" newcontext="true" ontimer="#Choices"
title="The Lottery Results">
       <timer value="50"/>
```

```
    <p align="center">
      <img src="lotlogo.wbmp" alt="Lottery"/>
      <br/>
      <br/>
      <b>The Lottery Results</b>
    </p>
  </card>
```

Starting the application

Now for that first Menu. In the interests of simplicity there is a simple list of links or anchors which point directly to appropriate cards in the deck.

To enable us to return to this card from any other card in the deck, the card id is set to 'choices'. In the meantime the simple list points to the id of the cards in the deck: 'Satlot1' is the card at the start of a series of cards marking up results from recent Saturdays, 'Wedlot1' does the same for Wednesday's results, and 'LuckyDip' refers to the card that invokes the random number routine.

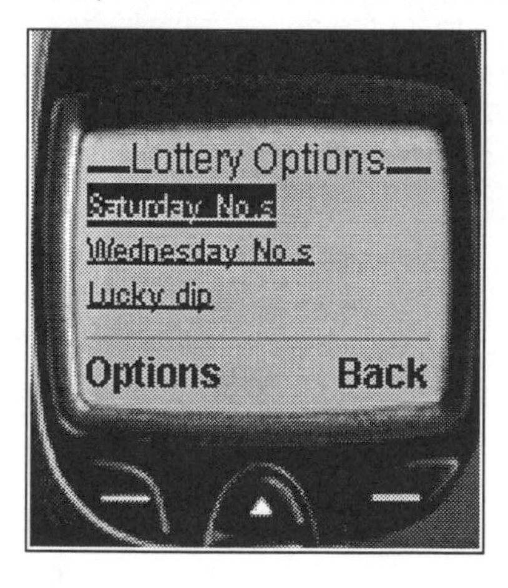

A more sophisticated approach might be to present an options list and act on the selection. For this application that is probably overkill so you can keep it simple. Note how the 'prev' and 'accept' soft keys are assigned functions by the template at the start of the deck

```
<!-- Then the Saturday Results -->
<card id="Choices" title="Lottery Options">
    <p><small>
       <a href="#Satlot1">Saturday No.s</a><br/>
       <a href="#Wedlot1">Wednesday No.s</a><br/>
       <a href="#LuckyDip">Lucky dip</a><br/>
    </small></p>
  </card>
```

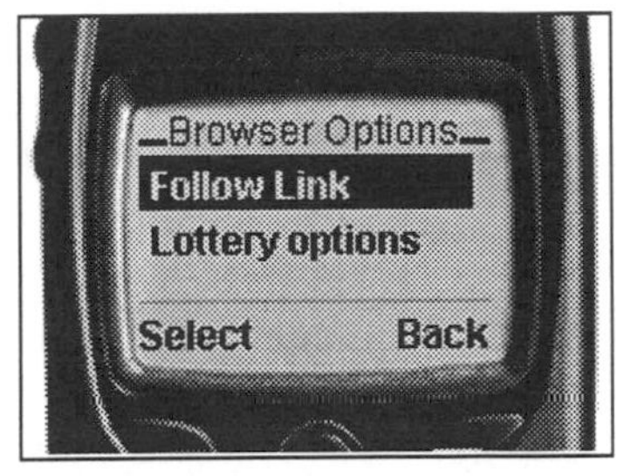

The different Browsers will render the options in various ways. In most cases, however, pressing select will either take you straight to the next card or a screen along the lines of the following example will be presentedand pressing 'select' again will take you to the next card. Scrolling down navigates the user back to the 'choices' card to choose another option

Delivering the content

Now you have to show the results for the chosen lottery day. Again in the interest of simplicity the results are placed in a straightforward chronological order - one set to a card.

The template at the start of the deck is shadowed to add the option to simply progress to the next card (imaginatively given the identifier 'Satlot2'). It is worth re-emphasising at this point that case sensitivity operates on references and links (unlike many browsers on PCs).

```
<card id="Satlot1" title="www.lottery.co.uk">
    <do type="accept" label="Next">
      <go href="#Satlot2"/>
    </do>
      <do type="accept" label="Lottery options">
            <go href="#Choices"/>
      </do>
    <p>
        <strong>Saturday </strong> <em><small> May 6
</small></em>
    </p>
    <p>
      <em><small> 6,11,19,38,39,40 </small></em>
    </p>
    <p>
      <em><small> Bonus Ball 2 </small></em>
    </p>
  </card>
```

Which presents itself as follows, note the layout is a 'best effort' to fit the test as required to the screen. During testing some emulators had a little trouble rendering even this simple text presentation. So to avoid unwanted breaks, etc., the formatting here is rather a brutal overkill. Usually the <br/> element is quite sufficient to display discrete lines of text.

```
  <card id="Satlot2" title="www.lottery.co.uk">
    <do type="accept" label="Next">
      <go href="#Satlot3"/>
    </do>
      <do type="accept" label="Lottery options">
            <go href="#Choices"/>
      </do>
    <p>
        <strong>Saturday </strong><em><small> Apr 29
</small></em>
    </p>
    <p>
      <em><small>
15,16,19,27,29,42 </small></em>
    </p>
    <p>
      <em><small> Bonus Ball 24 </small></em>
    </p>
  </card>
```

Again the options attached to the left hand soft key are displayed when it is first clicked. Note that now the code is 'shadowing' the template - that is it's overriding the original settings.

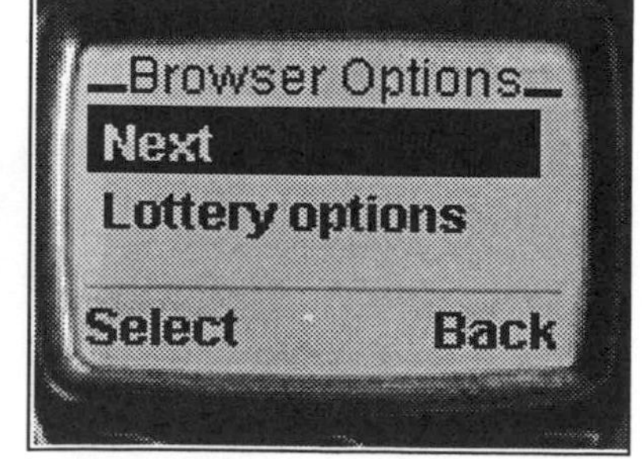

```
 <card id="Satlot3" title="www.lottery.co.uk">
    <do type="accept" label="Next">
      <go href="#Satlot4"/>
    </do>
      <do type="accept" label="Lottery options">
            <go href="#Choices"/>
      </do>
    <p>
        <strong>Saturday </strong><em><small> Apr 22
</small></em>
    </p>
    <p>
      <em><small>
26,31,35,43,48,49 </small></em>
    </p>
    <p>
      <em><small> Bonus Ball 38 </small></em>
    </p>
  </card>

  <card id="Satlot4" title="www.lottery.co.uk">
      <do type="accept" label="Lottery options">
            <go href="#Choices"/>
      </do>
    <p>
        <strong>Saturday </strong><em><small> Apr 15
</small></em>
    </p>
    <p>
      <em><small>
2,23,28,31,46,48 </small></em>
   </p>
    <p>
      <em><small> Bonus Ball 42 </small></em>
    </p>
  </card>
```

Throughout the linked list of cards, traversing the list is achieved simply by moving on to the next card or jumping back to the Main

menu. Needless to say the final card in the list should only give the option to go back to the main menu.

Then the Wednesday Results

Now the Wednesday results need to be listed in just the same way (with equally interesting card ids) - this listing doesn't have the irritating screen shots so you can get a clearer view of how the cards link together

```
<!--And next the Wednesday numbers -->
<card id="Wedlot1" title="www.lottery.co.uk">
    <do type="accept" label="Next">
      <go href="#Wedlot2"/>
    </do>
      <do type="accept" label="Lottery options">
            <go href="#Choices"/>
      </do>
    <p>
       <strong>Wednesday </strong> <em><small> May 3
</small></em>
    </p>
    <p>
      <em><small> ?,?,?,?,?,? </small></em>
    </p>
    <p>
      <em><small> Bonus Ball ? </small></em>
    </p>
  </card>

  <card id="Wedlot2" title="www.lottery.co.uk">
    <do type="accept" label="Next">
      <go href="#Wedlot3"/>
    </do>
      <do type="accept" label="Lottery options">
            <go href="#Choices"/>
      </do>
```

```
    <p>
        <strong>Wednesday </strong><em><small> Apr 26
</small></em>
    </p>
    <p>
      <em><small>
4,9,14,28,36,39 </small></em>
    </p>
    <p>
      <em><small> Bonus Ball 6 </small></em>
    </p>
  </card>

  <card id="Wedlot3" title="www.lottery.co.uk">
    <do type="accept" label="Next">
      <go href="#Wedlot4"/>
    </do>
      <do type="accept" label="Lottery options">
            <go href="#Choices"/>
      </do>
    <p>
        <strong>Wednesday </strong><em><small> Apr 19
</small></em>
    </p>
    <p>
      <em><small>
1,3,7,19,29,44 </small></em>
    </p>
    <p>
      <em><small> Bonus Ball 25 </small></em>
    </p>
  </card>

  <card id="Wedlot4" title="www.lottery.co.uk">
    <do type="accept" label="Lottery Options">
      <go href="#Choices"/>
    </do>
    <p>
        <strong>Wednesday </strong><em><small> Apr 12
```

```
</small></em>
     </p>
     <p>
        <em><small>
12,22,26,34,19,9 </small></em>
     </p>
     <p>
        <em><small> Bonus Ball 44 </small></em>
     </p>
  </card>
```

Winning the Lottery

Finally, you add a utility to allow the user to receive six random numbers between 1 and 49. It's invoked from the main menu jumping to the card with the id attribute set to 'LuckyDip'. To get the variable initialised the 'onevent' element is used with the attribute 'type' set to 'onenterforward'. This results in the content of the element being invoked as the card is accessed. In this case the WMLScript file 'Lottery.wmls' is opened and the function LuckyDip invoked - passing the variable JustMaybe to be set.

```
<card id="LuckyDip" title="www.lottery.co.uk">
    <onevent type="onenterforward">
       <go href="Lottery.wmls#LuckyDip('JustMaybe')"/>
    </onevent>
    <p>
       <do type="accept" label="ReCalculate">
              <                    g                    o
href="Lottery.wmls#LuckyDip('JustMaybe')"/>
       </do>
       <do type="accept" label="Lottery options">
              <go href="#Choices"/>
       </do>
```

```
        <strong>Lucky Dip Suggests</strong>
        <br/>
              $(JustMaybe)

    </p>
  </card>
```

Of course the first set of numbers suggested might not suit the user, or perhaps there might be several tickets to fill in, so the card needs to offer a chance to generate another set of numbers. This is achieved by setting an option on the accept soft key to call the WMLScript routine again. The displaying of the variable is simply managed by using the $(variablename) as a placeholder in the card.

And that is about it, the deck needs to be terminated by the terminatlng tag of the `<WML>` element, `</WML>` as suggested at the beginning of the example.

Now all that's needed is a WMLScript function that produces the numbers.

Tying up Bits of String

The element `<go href="Lottery.wmls#LuckyDip('JustMaybe')"/>` in the final card invokes the function 'LuckyDip' in the compilation unit 'Lottery.wmls' which expects the parameter 'JustMaybe' -or more accurately the value of the variable JustMaybe to be passed to it. The function called is defined in the compilation unit 'Lottery.wmls' as being externally available, it might also be worth noting that the comments between `/**` and `*/` are ignored by the system as are remarks following // and the end of the line.

```
/**
 * Compile string with 6 Random numbers between 1 and
49
 */

/**
 * First Declare Variables
 */
```

All variables in the deck must be declared before use using the var statement. The variables for each operation are declared here for ease of use and to allow simple debugging when the inevitable errors arise (keep an eye on those semi-colons):

```
extern function LuckyDip(JustMaybe) {
    var Bugger=""; //debug variable
    var Nums="xxxxxxxxx";
    var First=Lang.random(49)+1;
    var Second=Lang.random(49)+1;
    var Third=Lang.random(49)+1;
    var Fourth=Lang.random(49)+1;
    var Fifth=Lang.random(49)+1;
    var Sixth=Lang.random(49)+1;

/**
 * Then generate the six unique Numbers
 */
```

Here the use of the while loop is neatly illustrated as each number is checked to ensure it's unique and recalculated if not:

```
    while (Second == First) {
       Second=Lang.random(49)+1;
    };
    while (Third == First || Third == Second) {
       Third=Lang.random(49)+1;
    };
    while (Fourth == First || Fourth == Second ||
Fourth == Third)    {
       Fourth=Lang.random(49)+1;
    };
    while (Fifth == First || Fifth == Second || Fifth
== Third || Fifth == Fourth) {
       Fifth=Lang.random(49)+1;
    };
    while (Sixth == First || Sixth == Second || Sixth
== Third || Sixth == Fourth || Sixth == Fifth) {
       Sixth=Lang.random(49)+1;
```

```
    };

/** test the string in the toolkit */
    Nums = First + " " + Second + " " + Third + " " +
Fourth + " " + Fifth + " " + Sixth ;
```

Having constructed the list of six unique numbers you want to keep things looking elegant by sorting them - conveniently illustrating how to use a function within WMLScript

```
Nums = LotSort(First,Second,Third,Fourth,Fifth,Sixth);
    WMLBrowser.setVar(JustMaybe, Nums); // Set the
value back at the browser
    WMLBrowser.refresh(); //and update the screen

};
```

The function that does the sorting also has one or two interesting features, (assuming you don't look too closely at the rather clumsy logic) mainly the use of the for loop to check each of the six numbers in turn and swap them around if needed. The structure of the for loop is

```
for (start variable; test condition; increment (or
decrement))
{
block of repeated statements
}
```

and the if statement which follows the format

```
if (test=true){
      statement block
      };
```

There is an optional form using else

```
if ( Expression ) Statement else Statement
```

```
function    LotSort(First,Second,Third,Fourth,Fifth,
Sixth)
{
    // Sort Lottery suggestions
    var Sorted="" ;
    var SwapFlag=0 ;// Allow early breakout if sorted
    var SwapVal=0   ;     // Swap Variable
    var n=1 ; //Counter
    for ( n=1; n<6; n++) { //all sorted after five
passes - max

        if (First > Second){
                SwapFlag=1 ;// at least one more loop
                SwapVal=First;
                First=Second;
                Second=SwapVal;
        };
        if (Second > Third){
                SwapFlag=1 ;// at least one more loop
                SwapVal=Second;
                Second=Third;
                Third=SwapVal;
        };
        if (Third > Fourth){
                SwapFlag=1 ;// at least one more loop
                SwapVal=Third;
                Third=Fourth;
                Fourth=SwapVal;
        };
        if (Fourth > Fifth){
                SwapFlag=1 ;// at least one more loop
                SwapVal=Fourth;
                Fourth=Fifth;
                Fifth=SwapVal;
        };
        if (Fifth > Sixth){
                SwapFlag=1 ;// at least one more loop
                SwapVal=Fifth;
                Fifth=Sixth;
```

```
                Sixth=SwapVal;
        };

        if (!SwapFlag){
                        break ;
                        };

    };          //End for loop
    Sorted = First + " " + Second + " " + Third + " "
+ Fourth + " " + Fifth + " " + Sixth ;
    return Sorted;

};
```

It's also worth noting the use of return to send the sorted string back to the calling routine.

If the WML deck and the WMLScript are dutifully entered into lottery.wml and lottery.wmls respectively, these can be uploaded to a WAP enabled server (either locally using the Personal Web Server for NT or Windows, or one of the many WAP enabled ISP hosting services).

Pointing your WAP devices at http://yourdomain.your-ISP.co.uk/lottery.wml or whatever the appropriate URL might be, you should get the application running just like the sample you will find by pointing your WAP phone or emulator at:

http://www.net-works.co.uk/wapsamp1.wml

Don't forget, to save you typing in this example, you will find all the code on the web page: www.net-works.co.uk/wapsample1.htm

Cutting Lengths of String

If the results data was kept in a separate file, and read each time, then maintaining the results would be far easier and less prone to knadgering the WML by the slip of a keystroke. To do this a fixed format for the results is needed.

A simple solution is as follows in the format of date, numbers

```
22-04-2000, 31-49-43-48-26-35-(38)
19-04-2000, 01-19-29-03-07-44-(25)
15-04-2000, 07-23-28-33-46-48-(42)
```

```
12-04-2000, 12-22-26-34-19-09-(44)
08-04-2000, 18-49-46-47-23-30-(31)
05-04-2000, 03-48-02-10-40-07-(37)
```

In the interests of simplicity this should be saved in a text file called something like 'LotteryRes.txt' preferably in the same area as the WML and WMLScript. As WMLScript has a limited capacity to read in data the whole file must be read at once as a single string, then cut up into appropriate chunks. We could parse the data to identify the start, but again this would use up valuable space in the WAP client so a better solution is to use fixed length fields and great care to avoid indiscriminate trailing spaces etc. The string can then be confidently converted into the appropriate variables.

Firstly we need to modify the first card to read the data. This is done by invoking a WMLScript function written to read-in the data file and then cut it up, as follows.

```
<!-- First card is a logo -->
  <card id="Logo" newcontext="true" ontimer="#Choices"
title="The Lottery Results">

<onevent type="onenterforward">
<                             g                              o
href="Stringtest3.wmls#ReadText('TextString','SubStrin
g1','SubString2','SubString3','SubString4','SubString5
','SubString6','DateString1','DateString2','DateString
3','DateString4','DateString5','DateString6','Lucky1',
'Lucky2','Lucky3','Lucky4','Lucky5','Lucky6')"/>
       </onevent>
    <timer value="50"/>

      <p align="center">
        <img src="lotlogo.wbmp" alt="Lottery"/>
        <br/>
        <br/>
        <b>The Lottery Results</b>
      </p>
  </card>
```

The function to do the reading-in and cutting-up is unremarkable (apart from the appalling programming technique) except for the use of three different standard libraries. The file is read using the URL library function **url.loadstring('filepath')**, it is then cut using the string function **string.substring(target string, Start position, Number of characters)** remembering to start at position zero. Finally the WMLBrowser function **WMLBrowser.setvar (variablename,expression)** is used to send the values to the variables in the WML cards. If you look carefully you can see evidence of frustration while the parameters for this routine were tested and refined - can you spot them?

```
extern function
ReadText(TextString,SubString1,SubString2,
SubString3,SubString4,SubString5,SubString6,DateString
1,DateString2,DateString3,DateString4,DateString5,Date
String6,Lucky1,Lucky2,Lucky3,Lucky4,Lucky5,Lucky6)

  {
WMLBrowser.newContext();

var TestTextString=""; //Primary Data

TestTextString=URL.loadString("http://www.badstar.free
serve.co.uk/LotteryRes.txt","text/plain");
// <sigh> Don't forget the CHR$(13) at the end of each
line of the input file - </sigh>

//Date1=String.subString(TestTextString,0,10);
WMLBrowser.setVar(DateString1,
String.subString(TestTextString,0,10));
WMLBrowser.setVar(SubString1,
String.subString(TestTextString,12,17));
WMLBrowser.setVar(Lucky1,
String.subString(TestTextString,21,2));
WMLBrowser.setVar(DateString2,
String.subString(TestTextString,36,10));
WMLBrowser.setVar(SubString2,
String.subString(TestTextString,48,17));
```

```
WMLBrowser.setVar(Lucky2,
String.subString(TestTextString,67,2));
WMLBrowser.setVar(DateString3,
String.subString(TestTextString,72,10));
WMLBrowser.setVar(SubString3,
String.subString(TestTextString,84,17));
WMLBrowser.setVar(Lucky3,
String.subString(TestTextString,103,2));
WMLBrowser.setVar(DateString4,
String.subString(TestTextString,108,10));
WMLBrowser.setVar(SubString4,
String.subString(TestTextString,120,17));
WMLBrowser.setVar(Lucky4,
String.subString(TestTextString,139,2));
WMLBrowser.setVar(DateString5,
String.subString(TestTextString,144,10));
WMLBrowser.setVar(SubString5,
String.subString(TestTextString,156,17));
WMLBrowser.setVar(Lucky5,
String.subString(TestTextString,175,2));
WMLBrowser.setVar(DateString6,
String.subString(TestTextString,180,10));
WMLBrowser.setVar(SubString6,
String.subString(TestTextString,192,17));
WMLBrowser.setVar(Lucky6,
String.subString(TestTextString,211,2));
WMLBrowser.refresh();
};
```

The Nokia toolkit allows us to see this clearly on the **Variables Tab**
To use the variables in the deck, each card needs to be modified to take advantage of the new information as follows. The new WML shows the **'DateStringx'**, **'SubStringx'** and **'Luckyx'** variables being expressed, but take care over the case sensitivity of these, as the WAP environment is totally unforgiving.

```
<card id="Satlot1" title="www.lottery.co.uk">
            <do type="accept" label="Next">
      <go href="#Satlot2"/>
```

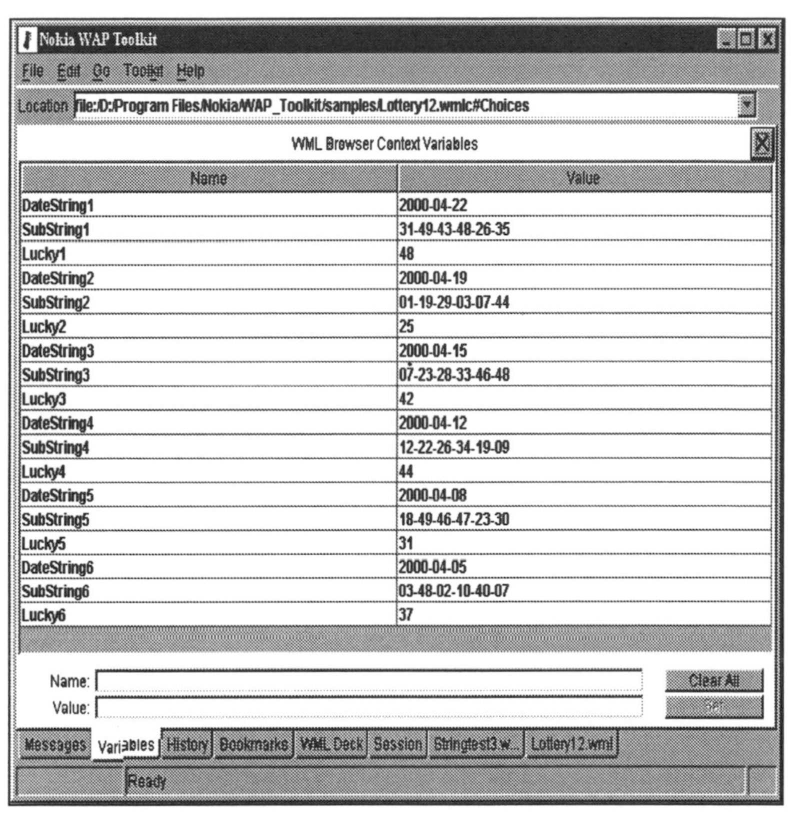


```
                </do>
        <do type="accept" label="Lottery options">
                <go href="#Choices"/>
        </do>
        <p>
        <strong>Saturday    </strong>    <em><small>
$(DateString1) </small></em>
                </p>
                <p>
        <em><small> $(SubString1) </small></em>
                </p>
                <p>
        <em><small> Bonus Ball $(Lucky1) </small></em>
                </p>
  </card>
```

and while you're tidying things up, have a look at using options rather than anchors in the opening menu using the following snippet:

```
  Select option
    <select name='LotteryType' ivalue='0'>
      <option onpick="#Satlot1">Saturday No.s</option>
      <option onpick="#Wedlot1">Wednesday No.s</option>
      <option onpick="#LuckyDip">Lucky dip</option>
    </select>
</p>
```

The initial screen, which when 'edit selection' is chosen, gives the screen...

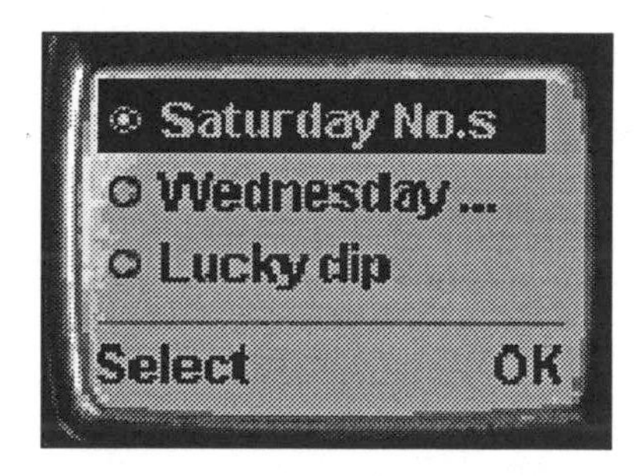

and on making a selection...

Was it all worth it? In fact whenever a WAP site is being developed, tradeoffs are made in development time, ease of use, bandwidth and the degree of device independence provided by the chosen code. Until the standards mature and the device manufacturers begin to conform without compromise to the 'spirit' of the standards, an enormous investment in testing and tweaking will be needed. It is safe to assume that it won't be too long before optimisation will only be needed for one or two broad device types - similar to the Explorer and Navigator considerations facing more conventional Internet developers.

Telephony Applications

There are ranges of programming topics which, although worth mentioning, are beyond the scope of this book, such as accessing the built-in features of the device. Known as telephony applications, the techniques used to do things like dial out through the telephone are documented by the WAP forum in specifications addressing what is known as the Wireless Telephony Application Interface (WTAI). Similar to the standard libraries they are invoked on devices that support them by prefixing the call appropriately from WML or WMLScript. There are a core of Telephony Applications of which one is the 'make call' which can be called by the WAP application as follows:

```
<WML>
<CARD>
<DO TYPE="ACCEPT" TASK="GO" URL="#eFood"/>
Welcome!
</CARD>
<CARD NAME="eFood">

<!--when the appropriate selection has been made make
the call -->
<DO                TYPE="ACCEPT"                TASK="GO"
URL="wtai://wp/mc;$FoodNum"/>
Choose Food:
<SELECT KEY="FoodNum">
<OPTION VALUE="5556789">Pizza</OPTION>
<OPTION VALUE="5551234">Chinese</OPTION>
<OPTION VALUE="5553344">Sandwich</OPTION>
<OPTION VALUE="5551122">Burger</OPTION>
<SELECT>
</CARD>
  </WML>
```

The `//wp/` defines the call as one of the 'core' or 'public' libraries. Other libraries include 'Voice call control', Network Text, and Miscellaneous.

There are two ways to invoke the function from WMLScript depending on how the author wants the user to initiate the dialing.

```
WTAPublic.makeCall("+18885551234");
    WTAVoiceCall.setup("+18885551234", true);
```

The various types of device functions are still in a state of flux as devices become ever more sophisticated. The working groups at the WAP forum have several "works in progress' and it's always a good idea to keep your finger on the pulse by checking the WAP forum web site for the current state of play.

Chapter 5

Getting Heavy - Advanced WAP

Please be warned, you are moving into a seriously techie acronym zone. The concepts are not difficult but, as always, the naming of the parts of the system is often so pretentious and long-winded (in the interests of clarity of course, not pomposity) that the need to abbreviate becomes very real.

Location Information

Convergence of Internet and wireless communication technologies is creating huge demand for access to Internet services from wireless handheld devices. This convergence creates a significant opportunity for telecommunications and Web portal providers to offer new and enhanced services for their customers who use handheld devices. It also gives telecommunications providers the opportunity to build a competitive advantage by deploying their own enterprise applications via handheld devices to mobile field workers. In both cases, the net result can be a more satisfied, loyal customer base.

Possibly the most exciting part of the WAP standards is the potential of the system to provide services based on where the device is located. Where is a restaurant near me? Where is the nearest ATM? Where is my next service call? Where is the cable that needs repair? Location information is where these applications add exceptional value to the wireless Internet and the providers will be transforming location-based information into relevant content for consumers, and into business advantage for organizations.

The current options all require a degree of input from the owner of the device. However, the coming generation of mobile devices ('phone' seems such an inadequate term) will provide 'always on' services with a far higher degree of accuracy when pinpointing the position of the device on the map. For example, BT Cellnet rolled out a General Packet Radio Service (GPRS) system into the UK in the summer 2000. Aimed at corporate customers, initially laptop users will be able to utilize the service via a plug-in card, but there will be full functionality phone handsets not far behind. Starting in the south of England, the deployment should be rapid enough to provide full coverage within a year.

The limitations of the cellular systems are principally competitive as getting the Telecoms companies to release cell information to service providers on the web could prove tricky. Along with the rather coarse granularity of the current system with some cells being perhaps 30Km across, it is entirely possible that some connection with the Global Positioning System (GPS) will be tied into the WAP devices in the near future. This has the advantage of providing a position accurate to less than 10m and being independent of the telco coverage in that area. It's not uncommon for petty squabbling amongst competitors to leave the road open for outsiders and there is every indication that location data will come from external services.

For the moment though, there is potential for such services within the cellular phone system with the WAP standard for GSM phones (the most common at the moment in Europe). The Wireless Telephony Application Interface specification defined by the WAP forum has functionality to provide information as follows...

Providing Location Information

This function is used to provide the current location information of the GSM terminal. The information available uniquely identifies the GSM cell in which the user is located at invocation time. The user must, at present, explicitly acknowledge the operation.

The WML instruction is invoked in a similar way to the Dialing function described in the earlier examples i.e.

```
URI: wtai://gsm/li [! <result>]
```

and to invoke the response in WMLScript the function is

```
WTAGSM.location
```

In both cases the return value is a string formatted as eight octets of the GSM location information in hexadecimal representation as follows:

Octets 1 - 3	Mobile Country & Network Codes (MCC & MNC)
Octets 4 - 5	Location Area Code (LAC)
Octets 6 - 7	Cell Identity Value (Cell ID)
Octet 8	Timing Advance

The mobile country code (MCC), the mobile network code (MNC), the location area code (LAC), the cell ID and the Timing Advance are coded as in GSM 04.08. In case of failure, the return value is a negative number and the WTAI error code.

The current range of location services will - in time - start to use one of these techniques to tell the user where the nearest curry house is without having to ask where the user to enter the local postcode (do you know the postcode of the street next to where you are, or the postcode of the street outside your cinema?). However a look at some of the current location based services is still worthwhile to show how things will evolve.

MultiMap

MultiMap.com provide both free and chargeable map information and the free service on the Internet is a very useful resource. For the current generation of WAP devices the service (probably correctly) declares the WAP screen as unsuitable for displaying map information. However the text-based service is still a fine application.

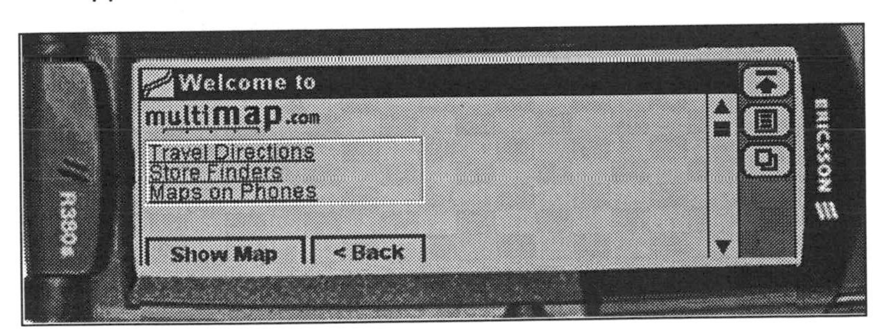

The first thing to notice is the different device has a totally different, but still WAP compliant, interface. The buttons are rendered on the screen rather than being set as programmed labels for soft keys. The whole layout for input is rather more sophisticated than that of a simple WAP telephone.

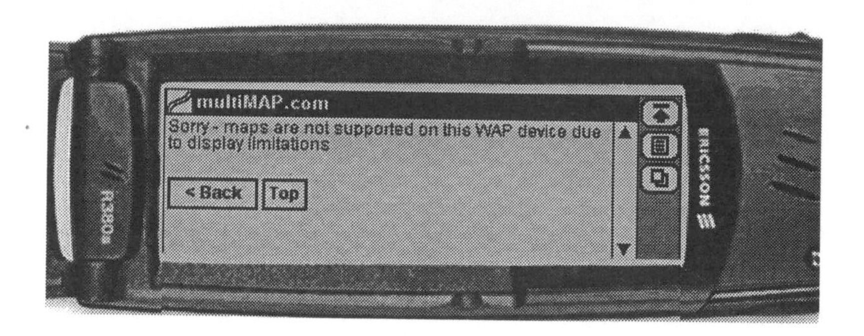

However the size of the display has little importance for text based applications as long as the scrolling mechanism allows information to be browsed. The directions provided are highly accurate and considering the service is free is as good a reason as any to use WAP on your next mobile phone. It also illustrates the way that, combined with 'always on' push services, accurate proximity information and good content WAP can soon challenge both current traffic information systems and any more imaginative ideas, such as payment for toll roads etc.

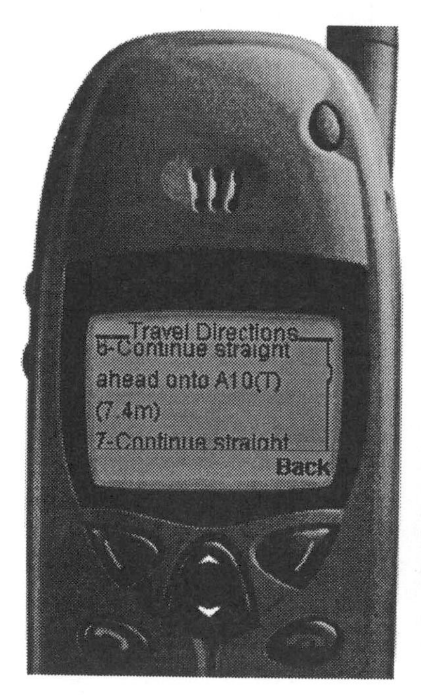

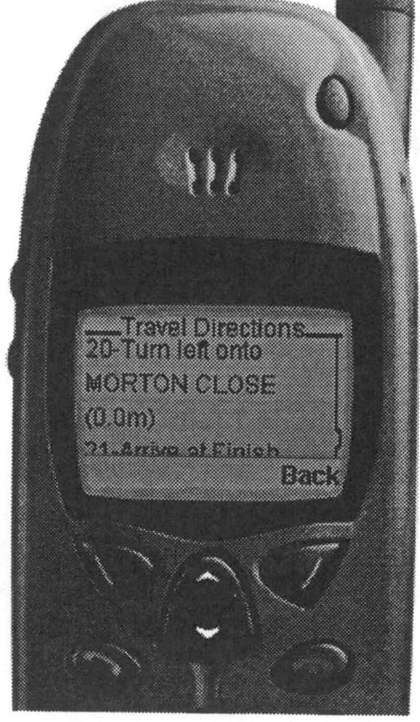

The other major player in Mapping information are the MapInfo Corporation who are developing some heavy duty commercial applications for various environments, such as workflow and location applications for PDAs as shown in the examples. They are also used for law enforcement software analysis and work is currently under way to move crime analysis and evidence collection onto small hand-held devices. When combined with the police's own communication network and powerful applications at the server this could be one of the major WAP applications.

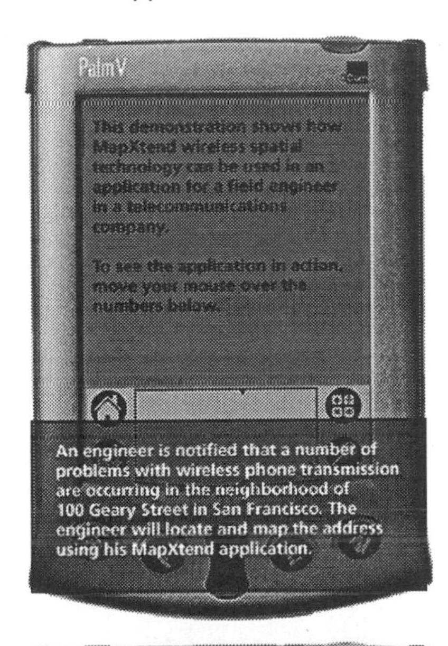

PalmV
This demonstration shows how MapXtend wireless spatial technology can be used in an application for a field engineer in a telecommunications company.
To see the application in action, move your mouse over the numbers below.
An engineer is notified that a number of problems with wireless phone transmission are occurring in the neighborhood of 100 Geary Street in San Francisco. The engineer will locate and map the address using his MapXtend application.

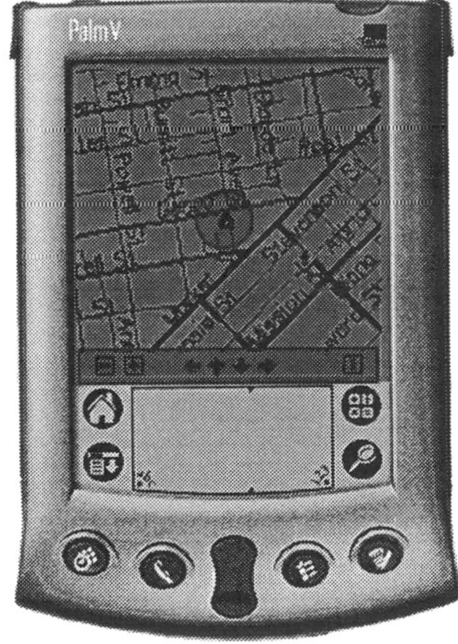

MapInfo Corporation invested in the Finnish company Karttakeskus Oy, a mapping technology and solutions company, giving MapInfo access to Karttakeskus' expertise in location-based WAP mobile applications, including its new Address Finder service based on MapInfo technology and developed with Nokia and Telia. With this investment, MapInfo will accelerate its own application of WAP technology.

The rationale for this was that Finland has the highest mobile phone usage per capita in Europe and is therefore a hotbed for new developments in the booming mobile voice and data industry, where MapInfo

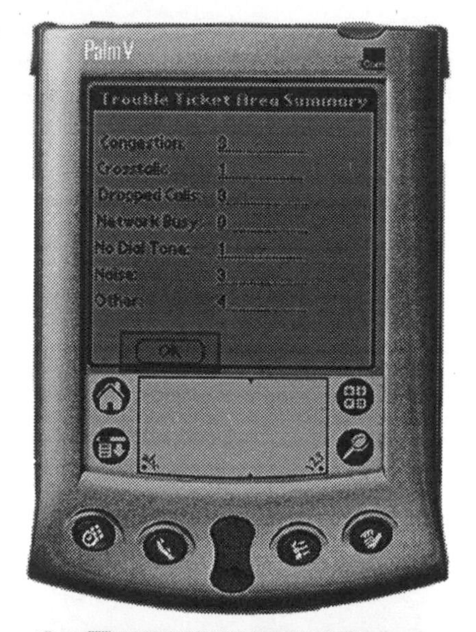

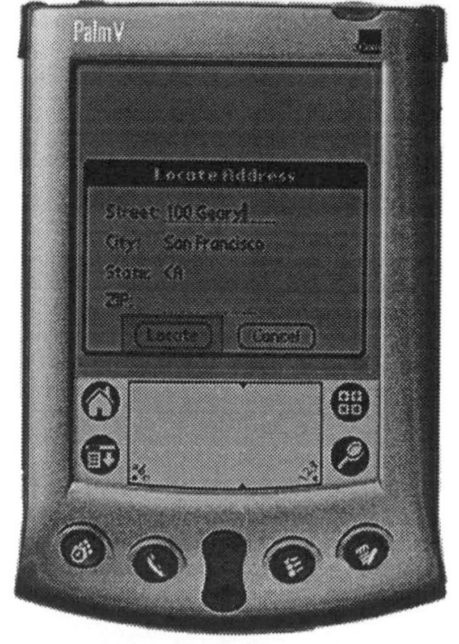

already maintains other business critical solutions. With the rapid growth in the spatial solutions market, MapInfo hope to be a world leader in this field.

The Address Finder service developed by Karttakeskus with Nokia and Telia uses MapInfo's MapXtreme Internet technology and enables mobile users to make location-based queries from mobile phones. The resulting map images are displayed on the phone's screen. Working with Nokia and Telia, Karttakeskus has pioneered previous online mapping applications, including Telia's InfoMedia Internet service as well as an address finder service for the first Nokia Communicator phones.

At the same time Motorola and Telespatial Inc. have joined forces with the effect that Telespatial will be part of Motorola's wireless Internet alliance program. Telespatial will offer real-time Location-Based Services (LBS) that will provide a direct, location-sensitive communication channel between brick-and-mortar retailers and Motorola mobile phone users.

Like the other services providers, the LBS will help wireless telephone users quickly find the nearest retailer offering the product or service they seek, such as a Petrol station, restaurant, retail store or ATM, it will provide travelers with directions to their hotel, business meeting location or recreational facility.

In fact WAP services offer a chance to get a slice of the expected $33 billion in business-to-consumer Internet sales back to high street businesses. These location-based services can direct mobile consumers to the nearest provider of the goods and services they are ready to purchase. Motorola like BT, Vodaphone or Cable and Wireless will offer this type of service to a rapidly growing base of WAP-enabled phone users, through a series of companion sites such as BT's Genie service.

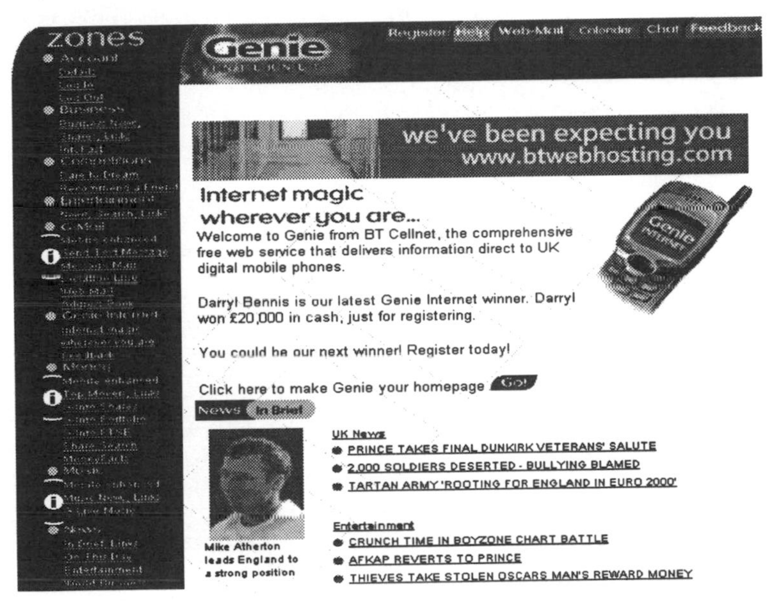

Ericsson and Palmtop Software have also unveiled map services for mobile devices as a range of location-based WAP services that will enable mobile access to local maps and pinpoint routes to selected destinations. The Ericsson range of services, called Location Based Information Services, will be available via operators and service providers on WAP phones as well as other devices supporting standards such as HTML. The LBIS claims to be designed with future technologies in mind and provide the network operator with a service-enabling platform for positioning applications over GPRS and 3G networks. Stored in a network operator's server, these services download compressed inform-ation upon request to any WAP-enabled phone or smartphone.

Pushing Yourself

The concept of "Push" in Internet terms is a mechanism where a system can transmit information to a client device without the browsing device client (the software on the device) making a request. In the normal course of events the client requests a service from a server which then responds by the most appropriate action, transmitting the results of that service request to the client. Known as pull technology this tends to underpin most of the World Wide Web and, indeed, the entire Client/Server networking model (the client is seen to pull the requested information from a comparatively passive server). In the Web example, the Browser transmits a URL to the web server which respond by sending the most appropriate web page. This can, of course, be either a simple static page, an active page of some sort (an application) or an error message (404 -page not found, springs to mind!).

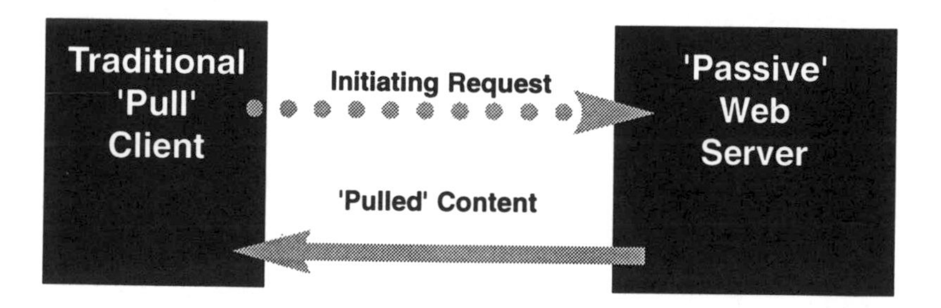

Push technology however, while still using the concept of a client and server, allows for a more "assertive" server which doesn't wait for a client request before transmitting its pages. Rather it makes a the decision to 'send' on some other basis such as a timed event or the status of a table on a database. In the case of future WAP developments even the location of the WAP device could be monitored and used to trigger Push activity.

In terms of the WAP world, a push happens when a Push Initiator (PI) transmits its content to a Client using either "Push-Over-The-Air" (OTA) protocol or the Push Access Protocol depending on what part of the end-to-end link this might entail.

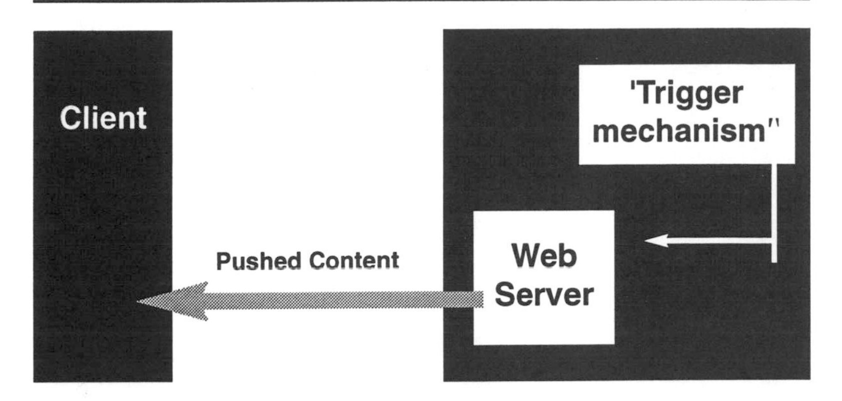

The dependency of the WAP client on a gateway means that there is no direct linking protocol between an Internet-based push initiator and the WAP client. However the gateway concept comes to the rescue in the form of a special gateway known as a Push Proxy Gateway (PGG) acting as an intermediary between the two ends of the connection.

The PGG accepts the contact from the PI server which then transmits its content. The PGG then takes the appropriate action to forward the "pushed" content to the WAP domain where the final delivery over the air takes place. The PGG also takes the role of notifying the PI about how successful the operation was, perhaps by waiting to see if the client accepts or rejects the content. It might even provide the PI with information about the client's capabilities (see 'User Agent Profiles') allowing the transmitted content to be tailored to suit the client.

The protocol used in the Internet side has been defined as the Push Access Protocol (PAP) and the WAP (OTA) side is managed by the Push-Over-The-Air Protocol.

The connection thus uses PAP to send XML messages (often tunnelling through familiar Internet protocols such as HTTP). The OTA protocol is based on the WSP services.

The Push Proxy Gateway (PPG) carries most of the burden in the push scenario, acting as it does, as the access point for content from the PI to the mobile networks. Its responsibilities therefore include security, client control and authentication, etc. The PPG controller therefore decides how 'push' can be implemented.

Concerns might be raised as to who can push content, who is allowed to receive what and under what circumstances. Clearly

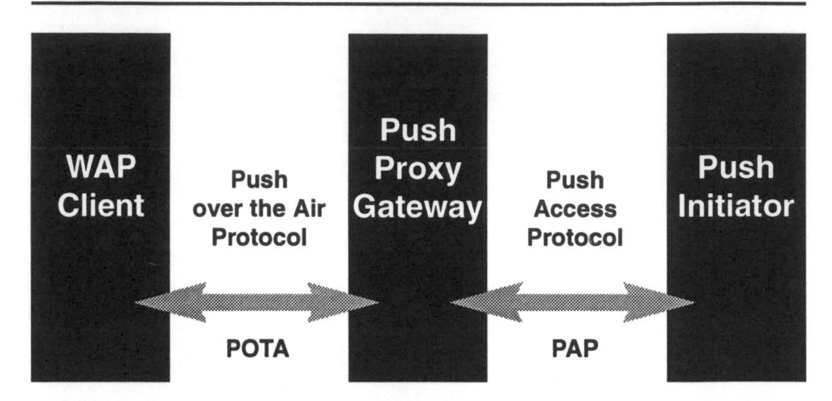

this is not a trivial task. Especially when the PPG functionality might be added to the conventional (pull?) WAP gateway where the benefits of sharing both types of service might be wanted.

Push Model Services

As the arbitrator of the whole push transaction, the PPG needs to be able to provide all the component services that the push model requires. From the PI side these are a service to rigorously identify the PI server and manage its access to the PPG system, parse the content control details, provide the client discovery services, resolve the addresses involved, carry out the recognised binary encoding inherent in the WAP definitions and the appropriate protocol conversion.

Content from the Internet is provided to the PPG using PAP, divided into various parts using a multipart/related content type. The first part of the message contains information for the PPG such as recipient information, call-back requests and so on. The PPG then responds to requests after parsing this control information by flagging the success of the operation, along with any other requested information it can derive.

After the content has covered this first hurdle and is recognised as a valid message, the PGG tries to find the destination device and deliver the content using the Push OTA Protocol. The attempt to push the message will continue until some timeout parameter is passed, having been set either by the PI server or the mobile operator.

Wherever possible, the WAP content types such as WML, WMLScript will be encoded into the denser binary formats allowing more effective use of the available bandwidth. Obviously this is best done before the over-the-air stage. If the PI has sent precompiled information then the PPG will recognise this and another boost to the performance can be gained. If special services such as broadcast operation or some form of multicast activity is required, (for example a penalty for England against Germany for a subscription service) an alias addressing scheme can be implemented by the PPG again extending the range of possible services on offer.

Tailoring Information for the Client

A PI server would probably start the push process by querying the PPG for any CC/PP information available (see UAPROF and PushPPG) to ensure that a suitable layout and activity content was presented for the client.
By definition PAP supports the following operations:

- Push Submission (Initiator to PPG)
- Result Notification (PPG to Initiator)
- Push Cancellation (Initiator to PPG)
- Status Query (Initiator to PPG)
- Client Capabilities Query (Initiator to PPG)

The Push Access Protocol has an XML style control entity, a content entity and an optional capability entity as a multipart/related message. The Control entity contains the delivery instructions for the PPG, whereas the content entity is the body of the message itself which may or may not be suitable for conversion by the PPG before transmission. The content entity might be included to indicate what the PI thinks the capability of the client is. Should the PI have requested confirmation then this is sent back via PAP depending to some extent on the bearer type. It will confirm that the message has either been sent or that it has been successfully delivered.

The PI implements push cancellation simply as an XML entity and the PPG flags either a success or failure. In a similar way the PI can interrogate the PPG regarding the status of the previously sent information with the obvious benefit of allowing the server to decide if a resend or alternative action should be taken.

Rather more intriguingly the Push Initiator can query the capabilities of the final client. This information is returned (where possible at least - older devices may not support the feature) as a partial presentation of the User Agent Profile in a standardised format allowing the server to adjust the content accordingly. With the increasingly common generation of dynamic content and markup this will become an essential part of managing output to WAP devices.

One of the more powerful features of WAP push is the ability to provide content to users asynchronously. In its basic form this is the ability to notify clients that there may be "information to their advantage" (to quote HG Wells) such as an important news item, a goal for (or against, in my case) the users favourite football team, or perhaps that the price of babyfood at the local supermarket is currently discounted by 9%.

The point of this is that accurate targeting of advertising or services can be pushed to users who don't have a direct service currently open. The Service Indicator (SI) content type at it's most simplistic consists of nothing more than a short identifying message and a URI pointing to the service. The end user can then chose to select the service or defer it by storing the details on the local client.

Staying Connected

The Push Over-The-Air Protocol acts as a layer over WSP and, as it name implies, acts between the PPG and the client devices. Normally using the WSP session as the delivery mechanism, a connection based push needs an active WSP session from the start as the server cannot establish the session. However the use of push when there is no active session is one of the most potentially useful features of WAP. So a mechanism has been developed to address this issue.

When there is a push to a device with no WSP session, a Session Initiation Application running on the client, with the task of listening to requests from the Over The Air server, springs into action deciding if setting up a push WSP session is safe. After checking the validity of the OTA server the Session Initiation Application, a WSP session is established.

And Finally - the Client

The most critical component on the Client is the Session Initiation Application (SIA). As mentioned previously, the connection-based 'push' needs to run over an established WSP session so as the client will, in general, not have a suitable session active with the push server - the SIA sits in the background listening for any push requests.

The request relayed by a PPG contains all the details needed to create the push session and the SIA interrogates this request. It then replies to the request by establishing a session with the PPG, indicating what the push options are (or simply pretends it hasn't heard if there is no suitable client capability or application) and opening up a suitable channel for content to pass through. The Client may also provide a generic wildcard application that should accept any application type.

The client can pre-empt some of this, of course, by indicating a willingness to accept push content while setting up a normal pull session.

Having passed this hurdle the details of the content are handed to the "Application Despatcher" which parses the type of content and decides which application, if any, should receive the data. The Application Despatcher then notifies the PPG that the content has been supplied to the destination application successfully (or otherwise).

In the connectionless-push only one secure port or one insecure port is available for the connection, and all pushes of this type must connect to one of these two ports.

Pushing an Addressing Scheme

The basic form of a push address is:

```
WAPPUSH=Address /type=userID@ppg.mobile.net
```

The type switch (which is always present) indicates the type of address. Addressing may be aceived on a device level, a user level and an application level. Of these three, the former two use the address-with-switch syntax, whereas the application-level addressing is a bit more complex, using a special application ID. The ppg.mobile.net part is the Internet host name of the Push Proxy Gateway.

Beyond this, the PI might need to target a specific user agent in the WAP device. This is achieved by using the application identifier such as X-Wap-Application-Id. This identifier, as with so many of the components in WAP, can be condensed from a URI into a token or numeric value in the interests of bandwidth expediency. The idea of 'well-known' user agents, such as the WAE, is built into the WAP specification and WINA assigns these well-known applications a consistent number.

Although the PI should use the scheme the PPG can translate any URI that it recognises as a having a WINA assigned numeric identifier. If the PI suggests both types of address reference then the binary form takes precedence, and if the URI is a recognised type, its translation takes precedence over both supplied addresses. It is possible for the PI to suggest unassigned values but this is to be discouraged due to the risk of collisions - the option exists for development of new and unimplemented user agents.

Establishing a Push Session

Who Are You?

When a Push Initiator starts off the process of building a push session, clearly it is essential that there is some form of security and authentication. The spread of malicious code through the Internet community with incidents like the "I love you" worm has shown that indiscriminate push activity could do much more than simply saturate the networks with unwanted advertising. There needs to be, therefore, a mechanism for ensuring that the PI is verifiably who they say they are and another mechanism (especially in m-commerce) to protect the transmitted content from prying eyes.

As for authentication of the PI, there are several possible solutions based as always on the answers the Web (and wider Internet community) have implemented. The most common solution is probably the use of session-based certificates using TSL/SSL. This puts the onus on the PPG to maintain a secure policy with its network security etc.

Object-level certificates that operate from end to end of the connection will become more popular as client devices carry their own signature, so bypassing the PPG. It is also possible to use the simple HTTP user/password authentication which, while simple to

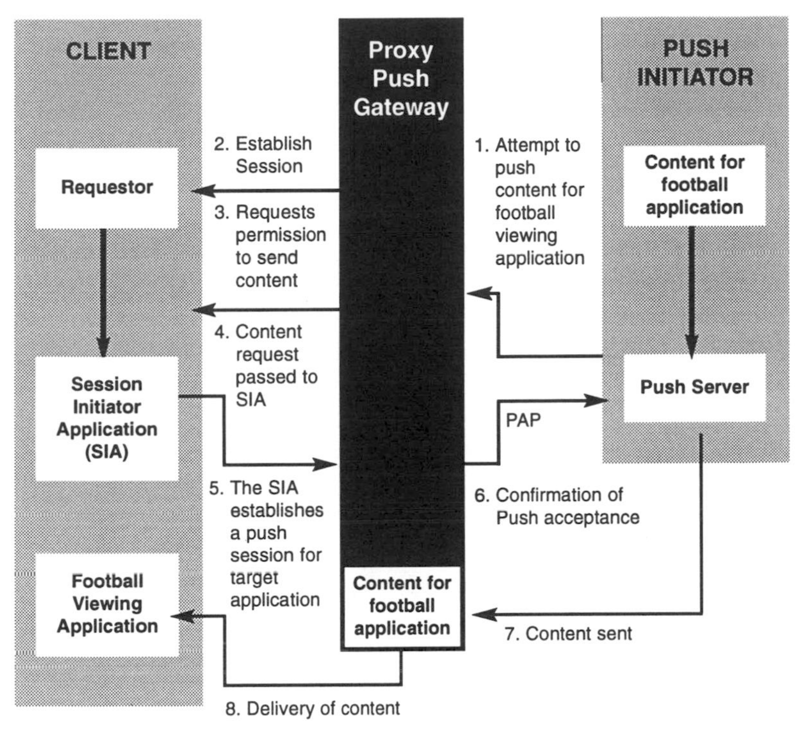

Establishing a Push session

implement, is not very secure and becomes difficult to manage when several sessions are involved.

While connected to the Internet each of these options (or a combination) is still theoretically open to some leakage, albeit only when exposed to huge number crunching capacity or desperately bad luck. The issue becomes far less critical if the WAP push-session comes from a private network - a situation likely to be quite common as intranets and specialist service providers emerge.

The User Agent Profile

One of the major differences between the major World Wide Web browsers running on computers and the WAP devices is the massive variation between the potential clients. These range from monochrome cell phones with as little as four lines of text to the smarter PDAs which can have quite reasonable screen resolutions (640x480) and 256 colours.

The challenge of presenting a useful output to all these devices requires some way of identifying the resources available to the client device and tailoring output accordingly. If use is to be made of the 'push' and 'location' possibilities of the recent WAP specifications then the server will need to know what the specifications of the client device are with respect to display, memory, browser type and any other parameter that either restricts the rendering of content or volunteers information about the user or her preferences.

Seeing as the World Wide Web Consortium has defined a consistent interface for device capabilities known as the Composite Capabilities/Preferences Profile, or the slightly catchier CC/PP, the WAP forum has joyfully embraced these options for the wireless equivalent, simply it seems, adding a few more acronyms to help maintain the mystique. This defines a system for describing the attributes of the browsing client using the standard attributes from what is known as the Resource Description Framework (RDF). For each device these attributes are grouped into collections of values/properties which can often override any default settings. Thus allowing the appropriate network component or server to address content specific needs.

On a lower level the design of the various protocol components (described in the first part of this book) has been optimised to allow effective profile caching by Web servers and proxies. By reducing the required bandwidth, the use of this device-specific information is made rather less painful (for the client at least) and the overhead is kept to a minimum.

In short, less waiting for better content - or, at least, more relevant and readable content.

The User Agent Profile or UAProf as it is known, (with the usual techie tendency to over-complicate this is also known as the "Capability and Preference Information" or CPI), extends the WAP standard to allow the end-to-end flow of the UAProf between the WAP client device, the routing network components and the source server.

Using the CCPP model to define the mechanism for describing and communicating CPI about the client, user and network for the WSP response, as mentioned earlier the system reduces bandwidth consumption with binary encoding and allowing efficient caching over a WSP session. The CPI can be extended where some form of translation is needed to tie up the source data and the presentation or processing abilities of the client.

As there is commercially useful information in the UAProf, it is to be expected that this specification will allow effective targeting of content based on the user's known preferences. The nature of the targeting remains to be seen but when added to proximity information the owners of the WAP gateways will have a tremendous capacity to offer useful services to people who might actually want them.

UAProf allows the system to capture various classes of device information including hardware and software definitions. However it is distinct from a User Preference Profile which would define a user's focused interests and requirements rather than the device being used. This is far more application-specific and would be defined by an application - such as a financial reporting service - rather than the WAP system.

The technology also allows cached profile queries to be serviced by WAP gateways while the client is off-line by third party requests and to designate a specific WSP session to target push operations to try and maintain the integrity of the profile. Low level WAP components (i.e. WTLS) are used to maintain the connection. However if the profile is invalidated in any way the source server should still try to transmit content as best it can.

At the same time mechanisms are currently being developed to allow activity to be resilient even when no active WSP session is in place, the client cannot be contacted or bandwidth is very limited.

To summarise. Attributes are collected on the client, encoded, transmitted (and cached) within a WSP session, potentially enhanced by specific requests for data, and network information.

The protocol used by push origin servers to retrieve the CPI from the WAP gateway or Push Protocol Proxy (PPG) is known snappily as WAP Push Access Protocol (PAP). The origin server makes the request over HTTP, and the response contains the profile, with MIME type text/xml for the system to operate on as necessary.

The end-to-end WAP system consists of five logical components:

- A client device capable of requesting and rendering WAP content.
- A wireless network employing WAP 1.1 or later protocols.
- A WAP gateway capable of translating WAP protocol requests into corresponding requests over the Internet and translating responses from the Internet into responses over the WAP protocols.
- The Internet or an intranet using TCP/IP-based protocols and possibly having one or more protocol gateways and Web/HTTP proxies.
- An origin (Web) server that can generate requested content. Though this specification refers to five end-to-end system components, actual configurations may physically deploy those components in many forms. For example, the latter three components (WAP gateway, Internet/intranet, and origin server) might easily be merged into a single server-side system connected to the WAP network. Moreover, the WAP gateway may itself be distributed, with different hosts serving as endpoints for different layers of the WAP protocol stack.

Of course the various physical parts - WAP gateway, Internet/intranet, and origin server - might easily be merged into a single system located on the gateway server connected to the WAP network. Alternatively, the WAP gateway may itself be distributed, with different hosts or servers serving as endpoints for different layers of the WAP protocol stack.

The CPI of the device is defined as only applying (scoped) to the particular WSP session within which it is transmitted. So each of a client's active WSP sessions may be associated with a different CPI tailored to be meaningful to the applications running on that origin server. The client may also update the cached CPI for a WSP session at any time, thus allowing the client to respond to any changes in the configuration of the application. For example, the user details might be updated or the display mode could change thus changing the most suitable layout for content.

The gateway uses World Wide Web standards in the form of the HTTP Extension Framework to convey the CPI information within HTTP headers. This means any even moderately experienced web author will have little trouble requesting details of the client capabilities and interpreting the responses (even if, in the case of early models, it's a confirmation that UAProf issues are not implemented).

The gateway controlled by the network operator can extend the received CPI with additional data obtained from local databases or other sources available only to the gateway, such as a network Home Location Register (HLR) or other enhanced information from the service provider. The WAP gateway may also add information to the profile to override information provided by the requesting device/user. These overrides may, for example, reflect policies in place by the network or gateway operator.

Security concerns are managed by opening a WSP session. Thus the UAProf-aware client can convey the profile information using special profile headers within the WSP Connect request which takes place during the initial 'handshaking' between the client and the gateway. If the client does not receive the "OK" Profile-Warning header in the WSP Connect response, it assumes that the gateway does not support this UAProf specification and therefore that the CPI is not being cached by the WAP gateway. As mentioned earlier the client may update the active UAProf at any time. This is achieved by the client transmitting what is known as a WSP Session Resume message to the WAP gateway which will then flush the old details from it's store and replace them with the updated information.

To suspend a WSP session, the client initiates a standard (as defined by the interminable protocol definitions) WSP Suspend request and unsurprisingly a client resumes a suspended WSP session by initiating a standard WSP Resume request.

This structure, which is explained in excruciating detail in the technical references on the WAP forum web site, allows not only for a veritable explosion of acronyms but also a system with all the advantages of the land-based Web browsers while addressing the highly personal needs of the mobile user. It is not unreasonable to assume that the typical WAP user will have more focused needs; less geared to 'surfing' but aimed more towards communicating or acting on information or a service. Hence the user profile allows application providers to provide tailored content that is both relevant and proactively useful. The world of the Star Trek communicator is here. Add cameras and assorted probes (Raman spectroscopy for materials analysis perhaps) talking to a complex database and the Tricorder so beloved of Mr Spock is a reality.

Chapter 6

Predicting the Future

Initially the next wave of applications are going to be quite mundane extensions of ordinary Web applications. For example, buying flowers can be done easily now via the web, but a market exists for late night delivery from the guilt-ridden who forget a partner's birthday while coming home from the pub.

In some forms these applications exist already. Following the above is the marrying up together of proximity data with the commercial opportunities. The local supermarket could monitor your purchases and notify you of any interesting, highly targeted special offers as you walk or drive past. You could be informed of car parking spaces - and book them. Along the same level of sophistication, public transport services could keep you up to date on the optimal route to any given destination. It could base its estimates on the current state of the various modes of travel and record your progress, perhaps organising your ticket details (or whatever token method is invented) based on when and where you are. Simply walk onto a train and your account can be debited, get off at the airport allow the TravelNet to keep you up to speed on the next plane to your destination and debit you again (or refund if the promised service levels are not adhered to). No need for queuing, waiting around uncertain of what is going on along the length of your journey. Imagine being in Darkest West London and knowing for certain that the best way to get to Cambridge is a bus to High St Kensington then the underground to Liverpool St Station where the next train to King's Lynn stops at Cambridge. And that it will get you there 35 minutes faster than the "official" route via King's Cross. And you don't even need to miss the train because a confused alien is blocking the ticket queue.

The new and innovative designs of mobile cellular devices will change our perceptions of how we expect these devices to work. The walkman is an obvious target - downloading music to the

PDA/phone would allow the user to keep a selection of music somewhere on the net and access it as and when they wanted. The cascade down to the gamers' market allows the possibility of playing multi-user games for everything from chess to complex graphic-based arcade-style adventure games. Again the immediacy and simplicity of the delivery means that the only limit is the imagination.

As the power increases only minimally the client devices can take on more server attributes, cheap CCD Cameras can be incorporated allowing us to record ever more of our lives. Ericsson are currently working on the concept of the mobile conference with video - something bound to be snapped by busy executives en-route to the most popular golf course. Remote and local store combined with the 2Mbit data rates promised by the new services offer an almost unlimited list of opportunities for innovation. The personal digital assistant will be able to take minutes, automatically book tickets, and alert a user to potential difficulties to a strategic proposal. The tedious computer-based issues of Business Intelligence and Customer relations could be built into a PDA for an executive (or even the au pair) whenever they are away from the "office".

However the real future of mobile data is clearly much more exciting - as the clients become ever more sophisticated and the communications underlying the mobile web become faster and more robust, the opportunity for us to totally re-invent our social structure becomes apparent. There are almost no areas where the technology cannot be applied - some wonderful, some sinister. We can have our cars sending out messages to the recovery services with a clear indication of what is wrong, and children can have their freedom while the web ensures that they are not out of touch from their parents. Our purchasing and meeting habits can be refined and organised to allow us to keep pace with the confusion and turbulence that the Knowledge Age seems to be dragging us into.

The sinister side is the possibility that we allow the "authorities" to use all this information (probably in the name of 'public good') to monitor and control us. Already the Government wants access to our email. With the growth of the connected community this would allow an unscrupulous leadership to scan for any level of subversion or apathy. There are two hopes against this happening; firstly as people become more connected the opportunities to circumvent officialdom become more apparent - hence the recent lifting of the ban on encryption exports in the US (the ban simply hit US business and people went on finding other ways to hide information). The other is that the continual changes in the underlying technology mean that unless a government wants to lose it's competitive advantages it has to allow the communication to become freer faster and ever more "personal" - so reading everyone's mail becomes totally counter productive. It has also been pointed out that the unattributable pay-per-use phones are far more of a crime risk for the security and law enforcement agencies.

Although still in its infancy, WAP already has its problems, two members of the WAP forum Phone.com and Geoworks as squabbling about royalties on parts of the standard each

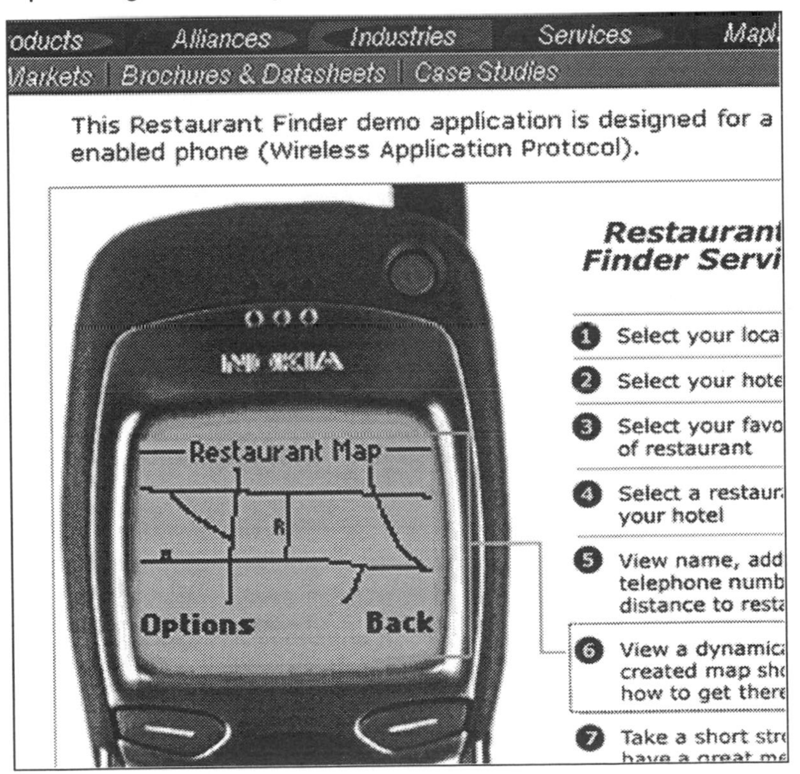

developed which tends to cause a certain amount of angst amongst would-be content providers and the thing most likely to slow thing up is a lack of real content. As the WAP model is enhanced and more exotic features implemented to distance it from its cousins such as SMS which isn't particularly interactive and has a limit of 160 characters per message (enough -you might think -from a phone pad) to add to this some later entrants are pushing more proprietary mobile solutions. In Japan NTTs iMode service is based on a cHTML or Compact HTML which is rather more similar to the familiar HTML. The take-up of iMode in Japan has been rapid and Logica are pushing a cHTML gateway called m-WorldGate. Whike these solutions will affect the way WAP evolves, little Web content is produced by hand-coding HTML sheets today and the battle will often be as to which environment has the best tools - at the moment WAP is a long way ahead.

Another factor to note is that the W3C are still evolving XML and a standard scripting format seems to be imminent. As XML becomes more prevalent throughout the WEB there seems an inevitability in the various mobile standards becoming drawn together as related subsets of XML implementations. Ultimately is it to be suspected that almost any device will happily chat to any other on - often with no human input at all.

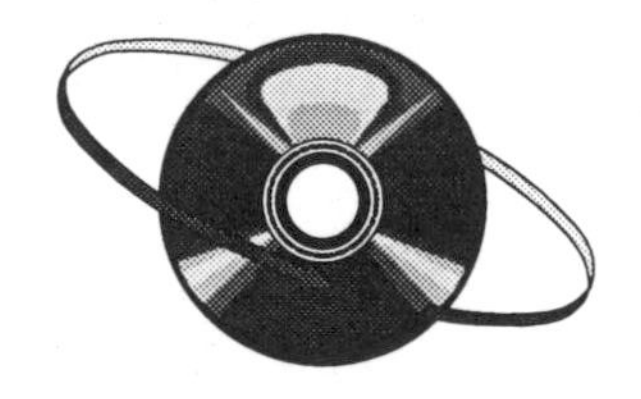

Chapter 7

What is 3G?

OK, so everyone has heard of 3G or third generation mobile phones. Apparently they are going to revolutionise our lives. But what is it, when is it coming and whatever happened to 1G and 2G?

A bit of history. The first generation mobile phones were those brick-sized affairs that weighed the same as an average car battery, but had a much shorter life. You'll recall the beach scene in the film Wall Street where Michael Douglas is showing off to the 'poor' Charlie Sheen. Actually, there are three faults with that scene. The first is that unless Mr. Douglas had been hitting the weights machine for months before hand, he would never have been able to hold the phone for that length of time. The second is that the battery would have run out halfway through the conversation, and the third is that he would never have got a signal.

And therein lay the problems with 1G. The analogue phones of the 80's were unreliable and, even if you did get a signal from the mast, you would more than likely find that the network was full or get cut-off in the middle of a call.

Enter the second generation of mobile phones. The 2G phones offer greater network efficiency by being digital. In short, they can cram more calls into the same bandwidth. Now, that would have been great if everyone had stayed making simple, short voice calls to one another. But since the space on the networks had been made available, the network operators started making 'value added' services a battleground for market share. In this batch of 'services' we received voicemail, text messaging and roaming capabilities (for which we needed dual-band phones), and latterly we've had Wap for connecting to the Internet.

Back to square one. Now the 2G networks are getting towards a state of congestion at peak periods and Internet connection speeds are slow. We need more bandwidth again.

In a nutshell, 3G is that extra bandwidth. You may have been aware of the bidding war that surrounded five new licences, which were made available by the Government. With no ceiling price set for each licence,

The Frequencies available in the 3G licences

Licence	(MHz)	(MHz)	(MHz)
Licence A	1920.0 - 1934.9	2110.3 - 2124.9	1914.9 - 1920.0
Licence B	1944.9 - 1959.7	2134.9 - 2149.7	
Licence C	1934.9 - 1944.9	2124.9 - 2134.9	1909.9 - 1914.9
Licence D	1959.7 - 1969.7	2149.7 - 2159.7	1899.9 - 1904.9
Licence E	1969.7 - 1979.7	2159.7 - 2169.7	1904.9 - 1909.9

and the only restriction being that Licence A had to go to a company other than BT Cellnet, Vodafone, Orange and One2-One, the total value came in at £22.47 billion. Tax cuts all round before the next election, and a major headache for the networks. They can obviously see a value in these licences, but how on earth are they going to make them pay?

Obviously there is more to 3G than the easing of network congestion. As the new bandwidth was used up by new services in 2G, so it will be in 3G. Current 2G networks allow Wap connections of around 9,600bps to 14.4kbps, but 3G will allow speeds in the region of 384kbps to 2,000kbps. Generic advertising has already begun with the promise of unlimited web browsing (the full HTML variation - graphics and all), interactive gaming and streaming-video capabilities.

Many in the media believe 3G is going to be the saviour of the human race and deliver miraculous, futuristic services within the next 12 months. Not so. Firstly there is nothing 'miraculous' about 3G. It is merely a mobilisation of what we have become accustomed to on our desktop PC's. The act of making the Internet accessible from a mobile phone handset will not automatically create a parallel raft of brilliant new technologies. And, secondly, 3G is not just round the corner.

Wa3Gons Roll

When will it all start? Now, that's harder to answer. The official date for UMTS service launch in Europe is on the 1st January 2002. Trials have already been held - successfully according to those involved, but then "they would say that wouldn't they".

The first planned commercial system is to be that run by Manx Telecom (owned by BT) on the Isle of Man. This is as much an experiment as a serious commercial venture, similar to the way in which the Tories tested the Poll Tax in Scotland before rolling it out throughout the rest of the UK.

By launching it in a restricted area they will hope to gain valuable experience in the construction and actual day-to-day running of the new generation before rollout. Thank you Isle of Man for de-bugging 3G on behalf of the rest of the country.

Even once the systems have been tested, there are still several hurdles to be overcome before 3G becomes a reality for the rest of us. The highest of these barriers is the require-ment for much smaller mobile network cells than those on which 2G currently runs.

Smaller cells, of course, means more masts. And more masts not only cost money but they need land. With the Government under a lot of pressure to stop selling off school land, and with health scares over electromagnetic radiation, finding places to stick new masts is not going to be a simple matter.

Add the cost of putting up all of these new masts to the cost of the original licence and you are talking very big money. Then you can add in the fact that whilst the UK may be an island, it is not the only country in the World. The networks will need to acquire 3G licences in all the other European countries in which mobile phones are used extensively. Mainland European Governments will have seen the huge sums paid in the UK, and you can imagine them already collectively saying, "We'll have some of that, thank you!"

So, the networks are not going to rush themselves into delivering 3G, at costs approaching something like the GDPs of smaller countries, when there is not a demonstrable demand. The most cost-effective solution is going to be a slowly-slowly approach, with the network operators gradually assessing which aspects of 3G are going to be the most profitable. The reason it all seems so imminent is the fact that the networks are all trying to gain a competitive advantage, before they have properly assessed the commerciality of such a strategy. Thus, in all probability, full coverage will not become a reality until 2004 or 2005.

Why Bother?

In two words: share price. To not participate in the rush toward 3G would dent the confidence of investors, causing the share price to fall. A falling share price would then have a knock-on effect on the competitiveness of the company - making the raising of new money more difficult and leaving the company open to the possibility of a hostile takeover.

The winners (or possibly the losers) in the bidding war

Winner	Licence	Amount	What they get
TIW	A	£4.38bn	The widest spectrum licence, reserved for a new entrant into the UK market. The real prize went to TIW backers Hutchinson-Whampoa (see Newsfeed), which owned 90 percent of the joint venture set up to get the licence.
Vodafone	B	£5.96bn	The widest spectrum available to an existing operator. Tactical bidding by BT forced Vodafone to pay a worryingly high price, and then they started legal action because Vodafone had been given more time to pay for the licence due to the sell-off of Orange.
BT	C	£4.03bn	A smaller capacity than license B - but nearly £2bn cheaper. A bargain at half the price!
One2One	D	£4.00bn	Has the same capacity as licence C. Deutsche Telekom, who own One2One will soon be taking part in the German mobile auction. They must have deep pockets.
Orange	E	£4.10bn	Also the same capacity as licence C. This licence helped Orange find a buyer in France Telecom.

However, there are more pressing commercial reasons than just a view of the stock market (though that didn't stop anyone, thank you lastminute.com). When the IDO and DDI networks in Japan introduced their IS-95B services, offering data speeds of up to 64kbps, there was a mass migration of subscribers over to them from NTT DoCoMo who had the slower network. That, in turn, forced NTT DoCoMo to reply with the launch of their i-mode service ahead of schedule.

Also, the rise in short message service use has given an insight into the possibility that innovative data services can be used to generate whole new revenue streams. In Finland, for example, who are streets ahead of the UK in terms of mobile usage, there was a 10-fold increase in SMS traffic during 1998. Mannesman, before being taken over, announced that their 100,000 subscribers spent more on SMS messages than they did on voice calls. And Sonera reported that 4% of their total revenues came from SMS traffic, which was close to 10% of its annual profits. The idea here is that new data services will emerge with the advent of 3G, making it all worthwhile.

But will customers, who are used to nearly-free Internet access at high speeds from their desktop, tolerate higher charges for the same information crammed onto a tiny mobile screen? It's my view that they will. As a race, we are becoming ever more impatient. When we want something, we want it now (hence all the negativity you are starting to see about Wap). So the fact that we can access the information we want, wherever we are, will cast aside our fears over the cost. Perhaps it will just be businesses who embrace 3G to start with, but that will trickle down to the average consumer. Remember the day when the person with the mobile phone was a high-flying businessman? Well, that was only a couple of years ago.

In summary, 3G is coming, it won't change your life spectacularly, it won't be here until 2004, and it will be pretty costly. The best option, in the meantime, is Wap - or should we call it 2G+ ?

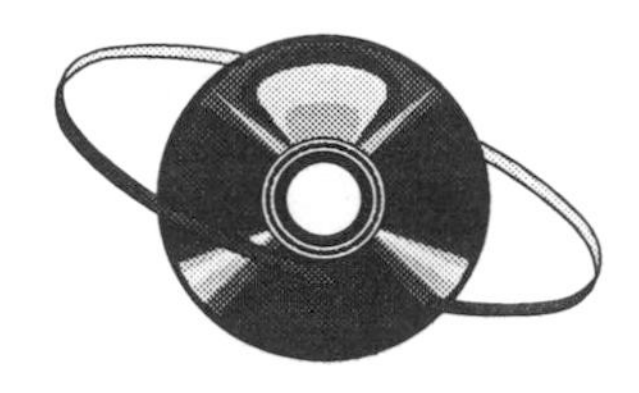

Appendix A

Major WML Elements and Attributes

Elements

<wml *attributelist >content* **</wml>**

Attribute:

xml:lang="*lang*"
Defines the language settings of the application

<card *attributelist> content* **</card>**

Attributes:

id="name"
identifies the card for navigation

title="*label*"
provides a titles for the WAP display

newcontext="*boolean*"
resets the browser context to just the current card and clears the variables

ordered="*boolean*"
Describes if the contents of the card should occur in a fixed sequence or not

onenterforward="*href*"
jump to the 'href' address or call the code found at that address

onenterbackward="*href*"
jump to the 'href' address or call the code found at that address

ontimer="*href*"
jump to the 'href' address or call the code found at that address

<template *attributelist> content* **</template>**

Attributes:

onenterforward="*href*"
see card attribute

onenterbackward="*href*"
see card attribute

ontimer="*href*">
see card attribute

<access *Attributelist/***>**

Attributes:

domain="*domain*"
Access is only permitted if the URL of the referring deck corresponds to the value of the attribute.

path="*path*"
Access is only permitted if the URL of the referring deck corresponds to the value of the attribute.

<meta *attributelist/***>**

Attributes:

content="*value*"
scheme="*format*"

```
<do attributelist>task</do>
```

Attributes:

type="*type*"
Defines the 'widget' attached to the task, options include:

accept	Positive acknowledgement (acceptance).
prev	Navigates backwards through history.
help	Request for help. May be context-sensitive.
reset	Clearing or resetting state.
options	Context-sensitive request for options or additional operations.
delete	Delete item or choice.

label="*label*"
Provides the displayed label for the 'widget'

name="*name*"
Identifies the name of the do event

optional="*boolean*">
Decides if the user agent may safely ignore this event

```
<onevent attributelist> task </onevent>
```

Attributes:

type="*type*"

onenterforward="href"
jump to the 'href' address or call the code found at that address

onenterbackward="href"
jump to the 'href' address or call the code found at that address

ontimer="href"
jump to the 'href' address or call the code found at that address

<postfield *attributelist***/>**

Attributes:

name="*value*"
Specifies the field name for transmission to an origin server

value="*value*"
Specifies the field value for transmission to an origin server

<go>*action***</go>**

Attributes:

href="*href*"
Target for the 'go' action

sendreferer="*boolean*"
Decides if the URL of the deck is sent to the requesting server allowing access control

method="*method*"
Declares if either the http 'post' or 'get' method is to be used

accept-charset="*charset*" >
Defines the character encoding to be used by the server

<prev> *content* **</prev>** -
go to the previous address

<refresh> *content* **</refresh>**
update the specified variables

<input *attributelist***/>**

Attributes:

name="*variable*"
specifies the name of the variable to be set by user input

title="*label*"
provides a title for the input element which can be used in the display of the element

type="*type*"
text or password

value="*default*"
sets a default value (only if the variable has not been previously set)

format="*mask*"
Defines the format of the possible user input - the options are...

A Allows any uppercase alphabetic or punctuation character, i.e, uppercase non-numeric character.

a Allows any lowercase alphabetic or punctuation character, i.e., lowercase non-numeric character.

N Allows any numeric character.

X Allows any uppercase character.

x Allows any lowercase character.

M Allows any character. The user agent may choose to assume that the character is uppercase for the purposes of simple data entry, but must allow entry of any character.

m Allows any character. The user agent may choose to assume that the character is lowercase for the purposes of simple data entry, but must allow entry of any character.

*f Allows any number of characters; f is one of the above format codes and specifies what kind of characters can be entered. Note that this format can only be specified once and must appear at the end of the format string.

nf Allows n characters where n is a number from 1 to 9; f is one of the above format codes (except *f) and specifies what kind of characters can be entered. Note that this format can only be specified once and must appear at the end of the format string.

\c Displays the next character, c, in the entry field. Allows escaping of the format codes as well as introducing non-formatting characters so they can be displayed in the entry area. Escaped characters are considered part of the input's value, and must be preserved by the user agent.

emptyok="boolean"
defines if an empty response can be accepted

size="n"
defines the width of he text input area displayed

maxlength="n"
defines the maximum length of the users input in characters

tabindex="n"/>
identifies the position of the current element in the tabbing sequence of the card

<select *attributelist* **>** *content***</select>**

Attributes:

title="*label*"
provides a title for the input element which can be used in the display of the element

multiple="*boolean*"
defines if multiple selections can be made

name="*variable*"
identifies the variable to be set by the selection

value="default"
Sets a default value (only if the variable has not been previously set)
iname="*index_var*"
establishes the variable holding the index value of the selection

ivalue="*default*"
sets a default value for the iname variable (only if the variable has not been previously set)

tabindex="*n*"
identifies the position of the current element in the tabbing sequence of the card

<option *attributelist>content***</option>**

Attributes:

title="*label*"
provides a title for the input element which can be used in the display of the element

value="*value*"
sets a default value (only if the variable has not been previously set)

onpick="*href*"
jump to the 'href' address or call the code found at that address

<optgroup *attributelist> content* **</optgroup>**

Attribute:

title="label"

provides a title for the input element which can be used in the display of the element

<fieldset *attributelist> content* **</fieldset>**

Attribute:

title="label"

provides a title for the input element which can be used in the display of the element

<a or **anchor** *Attributelist>* **href="href"** text **</a** or /**anchor>**

Attribute:

title="*label*"

provides a title for the input element which can be used in the display of the element

<img *attributelist/>*

Attributes:

alt="*text*"

Defines alternative text if the image is not rendered

src="*url*"

the address of the image

localsrc="*icon*"

defines a local alternative which if available is loaded in preference

align="alignment"

top, middle or bottom

height="n"

the height of the image as a percentage of available screen space

width="n"

the width of the image as a percentage of screen space

vspace="n"
defines the amount of space above and below the image

hspace="n"
defines the amount of space to the left or right of the image

<timer *attributelist"*/>

Attributes:

name="*variable*"
The name of the countdown variable

value="value
sets a default value (only if the variable has not been previously set)

<setvar *attributelist"*/>

Attributes:

name="*name*"
the name of the variable to be set

value="*value*"
the value to set the variable with

<p *attributelist*/>

Attributes:

align="*alignment*"
defines the alignment of the text within the paragraph, - 'left', 'right' or 'center'

mode="*wrapmode*"
wrap or nowrap

<table *attributelist*/>

Attributes:

title="*value*"
provides a title for the input element which can be used in the display of the element

align="*alignment*"
Center 'C', Left 'L', Right 'R'

columns="*number*"
Defines the number of columns in the table

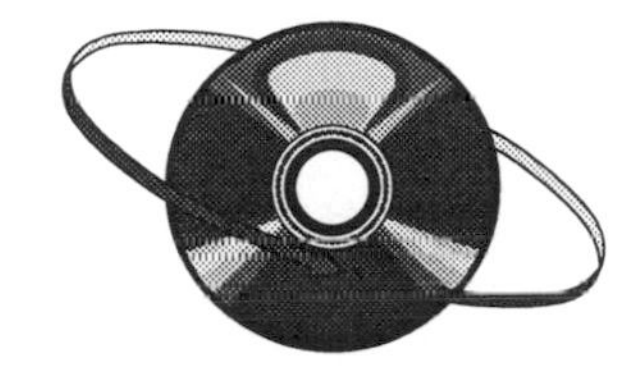

Appendix B

WMLS Standard Library Functions

Lang Library

Name: Lang

This library contains functions which support the core built in facilities of WMLScript.

abs

Usage: abs(value)

Returns the absolute value of the given number. The value returned corresponds to the calling data type of either floating-point or integer

Example:
```
var a = -5;var b = Lang.abs(a); // b = 5
```

min

Usage: min(value1, value2)

Returns the minimum value of the given two numbers. The value and type returned are the same as the value and type of the selected number. The selection follows WMLScript operator data type conversion rules. In the event of the values being equal, the first value is selected.

Example:
```
var a=-5;var b=Lang.abs(a);
var c=Lang.min(a,b); // c=-5
var d=Lang.min(45, 76.3); // d=45 (integer)
var e=Lang.min(45, 45.0); // e=45 (integer)
```

max

Usage: max(value1, value2)

Returns the maximum value of the two given numbers. The value and type returned are the same as the value and type of the selected number. The selection follows WMLScript operator data type conversion rules and if they are of the same value the first is selected

Example:
```
var a=-5;var b=Lang.abs(a);
var c=Lang.max(a,b); // c=5
var d=Lang.max(45.5, 77); // d=77 (integer)
var e=Lang.max(45.0, 45); // e=45.0 (float)
```

parseInt

Usage: parseInt(value)

Returns an integer value defined by the string value. Parsing the string until the first character is encountered that is not a leading + or - or a decimal digit.

Example:
```
var i=Lang.parseInt("12345"); // i=12345
var j=Lang.parseInt("10 m/s"); // j=10
```

parseFloat

Usage: parseFloat(value)

Returns a floating-point value defined by the string value. Parsing the string until the first character is encountered that cannot be parsed as being part of the floating-point representation.

Example:
```
var a=Lang.parseFloat("123.7"); // a=123.7
var b=Lang.parseFloat("+7.34e2 Hz"); // b=7.34e2
var c=Lang.parseFloat("70e-2 F"); // c=70.0e-2
var d=Lang.parseFloat("-1. C"); // d=-0.1
var e=Lang.parseFloat(" 100 "); // e=100.0
var f=Lang.parseFloat("Number:5.5"); // f=invalid
var g=Lang.parseFloat("7.3e meters); // g=invalid
var h=Lang.parseFloat("7.3e- m/s); // h=invalid
```

isInt

Usage: isInt(value)

Returns a boolean value that is true if the given value can be converted into an integer number by using parseInt(value). Otherwise false is returned.

Example:
```
var a=Lang.isInt("-123"); // true
var b=Lang.isInt("123.33); // true
var c=Lang.isInt("string"); // false
var d=Lang.isInt("#123"); // false
var e=Lang.isInt(invalid); // invalid
```

isFloat

Usage: isFloat(value)

Returns a boolean value that is true if the given value can be converted into a floating-point number using parseFloat(value). Otherwise false is returned.

Example:
```
var a=Lang.isFloat("-123"); // true
var b=Lang.isFloat("123.33"); // true
var c=Lang.isFloat("string"); // false
var d=Lang.isFloat("#123.33");// false
var e=Lang.isFloat(invalid); // invalid
```

maxInt

Usage: maxInt( )

Returns the maximum integer value.

Example:
```
var a=Lang.maxInt();
```

minInt

Usage: minInt( )

Returns the minimum integer value.

Example:
```
var a=Lang.minInt();
```

float

Usage: float( )

Returns true if floating-points are supported and false if not.

Example:
```
var floatsSupported = Lang.float();
```

exit

Usage: exit(value)

Ends the interpretation of the WMLScript routine and returns control to the caller of the WMLScript interpreter with the given return value. You can use this function to perform a normal exit from a function in cases where the execution of the WMLScript should be discontinued

Example:
```
Lang.exit("Value: " + myVal); // Returns a string
Lang.exit(invalid); // Returns invalid
```

abort

Usage: abort(errorDescription)

Aborts the interpretation of the WMLScript and returns control to the caller of the WMLScript interpreter with the return errorDescription. This function is used to perform an abnormal exit in cases where the execution of the WMLScript should be aborted due to serious errors detected by the calling function.

Example:
```
Lang.abort("Error: " + errVal); // Error value string
```

random

Usage: random(value)

Returns a pseudo-random positive integer value greater than or equal to 0 but less than or equal to the given value. If the value is a floating-point, Float.int( ) is first invoked to calculate the actual integer value.

Example:
```
var a=10;var b=Lang.random(5.1)*a; // b=0..50
var c=Lang.random("string"); // c=invalid
```

seed

Usage: seed(value)

Initializes the pseudo-random number sequence and returns an empty string.

Example:
```
var a=Lang.seed(123); // a=""
var b=Lang.random(20); // b=0..20
var c=Lang.seed("seed"); // c=invalid (random
seed// left unchanged)
```

characterSet

Usage: CharacterSet()

Returns the character set supported by the WMLScript Interpreter. The return value is an integer that denotes a value assigned by the Internet Address Numbering Authority for all character sets.

Example:
```
Var charset = Lang.characterSet(); // charset = 4
forlatin1
```

Float library

Name: Float

This library contains a set of typical arithmetical floating-point functions that are frequently used by applications for numeric analysis etc.

int

Usage: int(value)

Returns the integer part of the given value.

Example:
```
var a=3.14;var b=Float.int(a); // b=3
var c=Float.int(-2.8); // c=-2
```

floor

Usage: floor(value)

Returns the integer value that is nearest to but not greater than the given value

Example:
```
var a=3.14;var b=Float.floor(a); // b=3
var c=Float.floor(-2.8); // c=-3
```

ceil

Usage: ceil(value)

Returns the integer value that is nearest to but not less than the given value.

Example:
```
var a=3.14;var b=Float.ceil(a); // b=4
var c=Float.ceil(-2.8); // c=-2
```

pow

Usage: pow(value1, value2)

Returns an approximation of the result of raising value1 to the power of value2. If value1 is a negative number, value2 must be an integer.

Example:
```
var a=3;var b=Float.pow(a,2); // b=9
```

round

Usage: round(value)

Returns the integer number value that is closest to the given value. If values are equally close to the value, the result is the larger number value.

Example:
```
var a=Float.round(3.5); // a=4
var b=Float.round(-3.5); // b=-3
var c=Float.round(0.5); // c=1
var d=Float.round(-0.5); // d=0
```

sqrt

Usage: sqrt(value)

Returns an approximation of the square root of the given value.

Example:
```
var a=4;var b=Float.sqrt(a); // b=2.0
var c=Float.sqrt(5); // c=2.2360679775
```

maxFloat

Usage: maxFloat( )

Returns the maximum floating-point value supported by IEEE754 in single precision floating-point format.

Example:
```
var a=Float.maxFloat();
```

MinFloat

Usage: minFloat( )

Returns the smallest nonzero floating-point value supported by IEEE754 in single precision floating-point format.

Example:
```
var a=Float.minFloat();
```

String library

Name: String

This library contains a set of string functions, a string being an array of characters where each has an index. The first character (or element) in a string has an index of zero (0).

A white space character is one of the following characters:

- **TAB**: Horizontal Tabulation
- **VT**: Vertical Tabulation
- **FF**: Form Feed
- **SP**: Space
- **LF**: Line Feed
- **CR**: Carriage Return

Length

Usage: length(string)

Returns the length (number of characters) of the given string.

Example:
```
var a="ABC";var b=String.length(a); // b=3
var c=String.length(""); // c=0
var d=String.length(342); // d=3
```

IsEmpty

Usage: isEmpty(string)

Returns a boolean true if the string length is zero and a boolean false if the length is non-zero

Example:
```
var a="Hello";var b="";
var c=String.isEmpty(a); // c=false
var d=String.isEmpty(b); // d=true
var e=String.isEmpty(true); // e=false
```

charAt

Usage: charAt(string, index)

Returns a new string of length one containing the character at the position index of the string.

Example:
```
var a="My name is Joe";
var b=String.charAt(a, 0); // b="M"
var c=String.charAt(a, 100); // c=""
var d=String.charAt(34, 0); // d="3"
var e=String.charAt(a, "first") // e=invalid
```

subString

Usage: subString(string, startIndex, length)

Returns substring of the given string. The substring begins at position startIndex and its length is the given length. If the startIndex is less than 0, 0 is used for the startIndex. If the length is larger than the remaining number of characters, the length is replaced by the number of remaining characters. If the startIndex or the length is a floating-point, Float.int( ) is first used to calculate the actual integer value. If startIndex is larger than the last index or if length <= 0 an empty string ("") is returned.

Example:
```
var a="ABCD";var b=String.subString(a, 1, 2); //
b="BC"
var c=String.subString(a, 2, 5); // c="CD"
var d=String.subString(1234, 0, 2); // d="12"
```

find

Usage: find(string, subString)

Returns the index position of the first character in the string that matches the requested subString. If no match is found, the integer value -1 is returned. Two strings are defined to match when they are identical.

Example:
```
var a="abcde";
var b=String.find(a, "cd"); // b=2
var c=String.find(34.2, "de");// c=-1
var d=String.find(a, "gz"); // d= 1
var e=String.find(34, "3"); // e=0
```

replace

Usage: replace(string, oldSubString, newSubString)

Returns a new string resulting from replacing all occurrences of oldSubString in the given string with newSubString.

Example:
```
var a="Hello Jim. What is up Jim?";
var newName="Don";
var oldName="Jim";
var c=String.replace(a, oldName, newName);//
c="Hello Don. What is up Don?"
var d=String.replace(a, newName, oldName);//
d="Hello Jim. What is up Jim?"
```

Elements

Usage: elements(string, separator)

Returns the number of elements in the given string separated by the given separator. An empty string ("") is a valid element.

Example:
```
var a="My name is Joe; Age 50;";
var b=String.elements(a, " ");// b=6
var c=String.elements(a, ";");// c=3
var d=String.elements("", ";"); // d=1
var e=String.elements("a", ";"); // e=1
var f=String.elements(";",";"); // f=2
var g=String.elements(";;,;",";."); // g=4
separator=;
```

ElementAt

Usage: elementAt(string, index, separator)

Returns element of string at position index, the elements being separated by a separator, and returns the corresponding element. If the index is less than 0, the first element is returned. If the index is larger than the number of elements, the last element is returned. If the string is an empty string, an empty string is returned.

Example:
```
var a="My name is Joe; Age 50;";
var b=String.elementAt(a, 0, " "); // b="My"
var c=String.elementAt(a, 14, ";"); // c=""
var d=String.elementAt(a, 1, ";"); // d=" Age 50"
```

RemoveAt

Usage: removeAt(string, index, separator)

Returns a string where the element and the corresponding separator (if it exists) with the given index are removed from the given string. If the index is less than 0, the first element is removed. If the index is larger than the number of elements, the last element is removed. If the string is empty, the function returns a new empty string

Example:
```
var a="A A; B C D";var s=" ";
var b=String.removeAt(a, 1, s); // b="A B C D"
var c=String.removeAt(a, 0, ";"); // c=" B C D"
var d=String.removeAt(a, 14, ";"); // d="A A"
```

ReplaceAt

Usage: replaceAt(string, element, index, separator)

Returns a string with the current element at the specified index replaced by the given element. If the index is less than 0, the first element is replaced. If the index is larger than the number of elements, the last element is replaced. If the string is empty, the function returns a new string with the given element.

Example:
```
var a="B C; E";var s=" ";
var b=String.replaceAt(a, "A", 0, s); // b="A C; E"
var c=String.replaceAt(a, "F", 5, ";"); // c="B C;F"
```

InsertAt

Usage: insertAt(string, element, index, separator)

Returns a new string with the element and the corresponding separator (if needed) inserted at the specified element index of the original string. If the index is less than 0, 0 is used as the index. If the index is larger than the number of elements, the element is appended at the end of the string. If the string is empty, the function returns a new string with the given element.

Example:
```
var a="B C; E";var s=" ";
var b=String.insertAt(a, "A", 0, s); // b="A B C; E"
var c=String.insertAt(a, "X", 3, s); // c="B C; E X"
var d=String.insertAt(a, "D", 1, ";"); //d="B C;D;E"
var e=String.insertAt(a, "F", 5, ";");// e="B C; E;F"
```

Squeeze

Usage: squeeze(string)

Returns a string where all the consecutive series of white spaces within the string are reduced to one.

Example:
```
var a="Hello";
var b=" Bye Jon . See you! ";
var c=String.squeeze(a); // c="Hello"
var d=String.squeeze(b); // d=" Bye Jon . See you! "
```

Trim

Usage: trim(string)

Returns a string where all the trailing and leading white spaces in the given string have been trimmed.

Example:
```
var a="Hello";var b=" Bye Jon . See you! ";
```

```
var c=String.trim(a); // c="Hello"
var d=String.trim(b); // d="Bye Jon . See you!"
```

compare

Usage: compare(string1, string2)

The return value indicates the relationship of string1 to string2. The relation is based on the relationships between the character codes in the native character set. The return value is-1 if string1 is less than string2,0 if string1 is identical to string2 or 1 if string1 is greater than string2.

Example:

```
var a="Hello";var b="Hello";
var c=String.compare(a, b); // c=0
var d=String.compare("Bye", "Jon") // d=-1
var e=String.compare("Jon", "Bye") // e=1
```

toString

Usage: toString(value)

Returns a string representation of the given value. This function performs exactly the same conversions as WMLScript except that an invalid value returns the string "invalid".

Example:

```
var a=String.toString(12); // a="12"
var b=String.toString(true); // b="true"
```

Format

Usage: format(format, value)

Converts the given value to a string by applying the defined formatting information contained in the format string. The format specifier takes the following form: % [width] [.precision] type

The **width** argument is an integer controlling the minimum number of characters printed. If the number of characters in the output *value* is less than the specified width, the string is padded with spaces from the left until the minimum width is reached. The width argument never truncates value thus If the number of characters in the output value is greater than the specified width or if the width is not given, all characters of the value are printed (subject to the precision argument).

The **precision** argument also an integer, preceded by a period (.), that can be used to set the precision of the output value. The interpretation of this value depends on which of the following type identifiers is used:

d Defines the minimum number of digits to be printed. If the number of digits in the value is less than the precision value, the output value is padded on the left with zeroes. The value is not truncated when the number of digits exceeds the precision value. The default precision value is 1. If the precision value is specified as 0 and the value to be converted is 0, the result is an empty string.

f Defines the number of digits after the decimal point. If a decimal point appears, at least one digit must appear before it. The value is rounded to the appropriate number of digits with a default precision of 6 digits; if the precision is 0 or if the period (.) appears without a number following it, no decimal point is printed. When the number of digits after the decimal point in the value is less than the precision, it is padded with zeros to fill columns (e.g., result of String.format ("%2.3f", 1.2) will be " 1.200")

s Defines the maximum number of characters to be printed. By default, all the characters are printed. When the width is larger than precision, the width should be ignored. Unlike the width argument, the precision argument can cause either truncation of the output value or rounding of a floating-point value. The type argument is the only mandatory format argument and is specified after any optional format fields. The **type** character determines whether the given value is interpreted as integer, floating-point or string. The supported type arguments are:

d *Integer:* The output value has the form [-]nnnn, where nnnn is one or more decimal digits.

f *Floating-point:* The output value has the form [-] nnnn.nnnn, where nnnn is one or more decimal digits. The number of digits before the decimal point depends on the size of the target number, and the number of digits after the decimal point depends on the precision requested.

s *String:* Characters are printed up to the end of the string or until the precision value is reached. % can be presented by preceding it with another percent character (%%).

Example:

```
var a=45;
var b=-45;
var c="now";
var d=1.2345678;
var e=String.format("e: %6d", a); // e="e: 45"
var f=String.format("%6d", b); // f=" -45"
var g=String.format("%6.4d", a); // g=" 0045"
var h=String.format("%6.4d", b); // h=" -0045"
var i=String.format("Do it %s", c); // i="Do it now"
var j=String.format("%3f", d); // j="1.234567"
var k=String.format("%10.2f%%", d); // k=" 1.23%"
var l=String.format("%3f %2f.", d); //l="1.234567 ."
var m=String.format("%.0d", 0); // m=""
var n=String.format("%7d", "Int"); // n=invalid
var o=String.format("%s", true); // o="true"
```

URL library

Name: URL

This library contains a set of functions for handling and manipulating absolute and relative URLs and a collection of associated string handling functions. The general URL syntax supported is:

```
<scheme>://<host>:<port>/<path>;<params>?<query>#<fragment>
```

IsValid

Usage: isValid(url)

Returns true if the given url has a valid URL syntax, otherwise returns false. Both absolute and relative URLs are supported.

Example:

```
var a=URL.isValid("http://www.acme.com/script#func()");// a=true
var b=URL.isValid("../common#test()");// b=true
var c=URL.isValid("experimental?://www.acme.com/pub")// c=false
```

GetScheme

Usage: getScheme(url)

Returns the scheme used in the given url. Both absolute and relative URLs are supported.

Example:

```
var a=URL.getScheme("http://w.a.com"); // a="http"
var b=URL.getScheme("w.a.com"); // b=""
```

GetHost

Usage: getHost(url)

Returns the host specified in the given url. Both absolute and relative URLs are supported.

Example:

```
var a=URL.getHost("http://www.acme.com/pub");// a="www.acme.com"
var b=URL.getHost("path#frag");// b=""
```

GetPort

Usage: getPort(url)

Returns the port number specified in the given url. If no port is specified, an empty string is returned. Both absolute and relative URLs are supported.

Example:

```
var a=URL.getPort("http://www.acme.com:80/path");// a="80"
var b=URL.getPort("http://www.acme.com/path");// b=""
```

GetPath

Usage: getPath(url)

Returns the path specified in the given url. Both absolute and relative URLs are supported.

Examples:

```
var a=URL.getPath("http://w.a.com/home/sub/comp#frag");//
a="/home/sub/comp"
var b=URL.getPath("../home/sub/comp#frag");//
b="../home/sub/comp"
```

GetParameters

Usage: getParameters(url)

Returns the parameters used in the given url. If no parameters are specified an empty string is returned. Both absolute and relative URLs are supported.

Example:

```
var a=URL.getParameters("http://w.a.c/scr;3;2?x=1&y=3");//
a="3;2"
var b=URL.getParameters("../scr;3;2?x=1&y=3");// b="3;2"
```

GetQuery

Usage: getQuery(url)

Returns the query part specified in the given url. If no query part is specified an empty string is returned. Both absolute and relative URLs are supported.

Example:

```
var a=URL.getQuery("http://w.a.c/scr;3;2?x=1&y=3");//
a="x=1&y=3"
```

GetFragment

Usage: getFragment(url)

Returns the fragment used in the given url. If no fragment is specified an empty string is returned. Both absolute and relative URLs are supported.

Example:

```
var a=URL.getFragment("http://www.acme.com/cont#frag")
;// a="frag"
```

GetBase

Usage: getBase( )

Returns an absolute URL (without the fragment) of the current WMLScript compilation unit.

Example:

```
var a=URL.getBase();// Result: "http://www.acme.com/test.scr"
```

GetReferer

Usage: getReferer( )

Returns the smallest relative URL (relative to the base URL of the current compilation unit) to the resource that called the current compilation unit.

Example:

```
var base =URL.getBase();// base ="http://www.acme.com/
current.scr"
var referer =URL.getReferer();// referer ="app.wml"
```

Resolve

Usage: resolve(baseUrl, embeddedUrl)

Returns an absolute URL from the given baseUrl and the embeddedUrl according to the rules specified in RFC2396. If the embeddedUrl is already an absolute URL, the function returns it without modification.

Example:

```
var a=URL.resolve("http://www.foo.com/","foo.vcf");//
a="http://www.foo.com/foo.vcf"
```

EscapeString

Usage: escapeString(string)

This function returns a new version of a string value in which special characters are replaced by a hexadecimal escape sequence (you must use a two-digit escape sequence of the form %xx). The characters to be escaped are:

Control Characters: US - ASCII coded characters 00-1F and 7F

Space: US - ASCII coded character 20 hexadecimal

Reserved: ";" | "/" | "?" | ":" | "@" | "&" | "=" | "+" | "$" |","

Unwise: "{" | "}" | "|" | "\" | "^" | "[" | "]" | "`"Delims: "<" | ">" | "#" | "%" | <">The given string is escaped as such: no URL parsing is performed.

Example:

```
Vara=URL.escapeString("http://w.a.c/dck?x=\u007ef#crd");//
a="http%3a%2f%2fw.a.c%2fdck%3fx%3d%ef%23crd"
```

UnescapeString

Usage: unescapeString(string)

The unescape function returns a new version of a string value in which each escape sequence of the sort that might be introduced by the URL.escapeString( ) function is replaced by the character it represents.

The given string is unescaped as such; no URL parsing is performed.

Example:

```
var a="http%3a%2f%2fw.a.c%2fdck%3fx%3d12%23crd";
var b=URL.unescapeString(a);// b ="http://w.a.c/dck?x=12#crd
```

LoadString

Usage: loadString(url, contentType)

Returns the content denoted by the given absolute url and the contentType. The given content type is erroneous if it does not follow the following rules:·

- You can only specify one content type. The whole string must match with only one content type and you cannot have any extra leading or trailing spaces;
- The type must be text but the subtype can be anything;
- The type prefix must be "text/"

Example:

```
var myUrl="http://www.acme.com/vcards/myaddr.vcf";myCard=
URL.loadString(myUrl, "text/x-vcard");
```

WMLBrowser library

Name: WMLBrowser

This library contains functions which WMLScript can use to access the associated WML context.

Allowing the application to collect or post data from the user client.

getVar

Usage: getVar(name)

Returns the value of the variable name in the current browser context. If the variable does not exist, returns an empty string. The variable name must follow the syntax specified by the WML Specification.

Example:
```
var a=WMLBrowser.getVar("name");
// a="false" or whatever value the variable has.
```

SetVar

Usage: setVar(name, value)

Returns true if the variable name is successfully set to contain the given value in the current browser context. Otherwise returns false. The variable name and its value must follow the syntax specified by the WML Specification. The variable value must be legal XML CDATA.

Example:
```
var a=WMLBrowser.setVar("name", Brian); // a=true
```

go

Usage: go(url)

Specifies the content denoted by the given url to be loaded. This function has the same semantics as the GO task in WML. The content is loaded only after the WML browser resumes the control back from the WMLScript interpreter.

Example:
```
var        card="http://www.lottery.co.uk/loc/app.dck#start"
;WMLBrowser.go(card);
```

Prev

Usage: prev()

Signals the WML browser to go back to the previous WML

card. This function has the same semantics as the PREV task in WML. The previous card is loaded only after the WML browser resumes the control back from the WMLScript interpreter.

Example: `WMLBrowser.prev();`

newContext

Usage: newContext( )

Clears the current WML browser context. This function has the same semantics as the NEWCONTEXT attribute in WML.

Example: `WMLBrowser.newContext();`

GetCurrentCard

Usage: getCurrentCard( )

Returns the smallest absolute or relative URL specifying the card (if any) currently being processed by the WML browser.

Example: `var a=WMLBrowser.getCurrentCard();// a="deck#input"`

Refresh

Usage: refresh( )

Forces the WML browser to update its context. The user interface is then updated to reflect the updated context. This function has the same semantics as the REFRESH task in WML.

Example:

```
WMLBrowser.setVar("name", "Pete");
WMLBrowser.refresh();
```

Dialogs library

Name: Dialogs

This library contains a collection of typical user interface functions.

Prompt

Usage: prompt(message, defaultInput)

Prompts for user input and displays the supplied message. The defaultInput parameter contains the initial content for the user input. Returns the user input.

Example:

```
var a="27 Morton Close";var b=Dialogs.prompt("Address: ",a);
```

Confirm

Usage: confirm(message, ok, cancel)

Displays a request for confirmation using the given message and two reply alternatives: ok and cancel. Waits for the user to select one of the reply alternatives and returns true for ok and false for cancel.

Example:
```
function onAbort() {return Dialogs.confirm("Are you
absolutely sure?", "Yes", "Maybe");};
```

Alert

Usage: alert(message)

Displays the given message to the user and waits for the user to acknowledge.

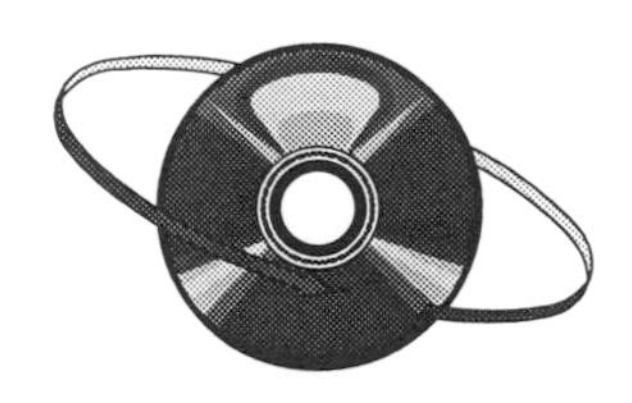

Appendix C

XML - eXtensible Mark-up Language

As the whole field is very much in its infancy, the "display" language for WAP known as the Wireless Markup Language (WML) is continually evolving. The WAP forum has committed to reusing existing standards as far as possible. To this end they have wisely chosen to base WML on the recently emerged standard for Internet content layout. Known as XML (for eXtensible Markup Language), this allows the creation of specialized tags that can be configured to meet a particular need. As a meta-language, XML underpins the whole structure of WML and an overview of how it functions will allow any further changes in the WAP standard to be more easily accommodated.

The Body that tries to manage the evolution of content delivery on the Internet is known as the World Wide Web Consortium or W3C. It perceived that, with the rapidly evolving content of the World Wide Web and indeed the whole role of the Internet in the modern world, the original method of displaying web pages - HTML (Hypertext Markup Language) - while hugely successful, was too limited. Its original source, the venerable Standard Generalized Markup Language (SGML), was far too bulky to be used on the light and dynamic applications that were beginning to appear so a search for a solution was begun by a W3C working group under Jon Bosak of Sun Microsystems.

XML has a structure similar to a standard Web page in that it is formed as a list of elements. These are a series of tags, which may contain inherent attributes that manage the associated data. As these elements are defined through Document Type Definitions and Stylesheets by the author of the XML dialect there are no correct or incorrect tags, as long as they are defined,

<Fluffy></Fluffy>, <Wetfont></Wetfont> and <Bruising></Bruising> can all be legitimate tags.

In XML, similarly to HTML, the tags are arranged in pairs (indeed rather more rigidly so as XML doesn't provide for the tolerance that many browsers grant poorly formed HTML).

The pairs have the format

```
<elementname attribute="value"> data </elementname>
```

although empty elements i.e.

```
<elementname attribute="value"></elementname>
```

are graciously permitted a condensed form

```
<elementname attribute="value"/>
```

Displaying XML

Data is held as tagged elements which are rendered by the browsing devices following rules laid down by the the Stylesheets and Document Type Definitions associated with this particular XML definition.

An example of a very simple XML document...Glassblade.xml

```
<?xml version="1.0" standalone="no"?>
```

This line is the XML declaration which informs the browser which version of XML is in use. The standalone element indicates that a separate file is needed for the DTD.

```
    <!DOCTYPE Glassblade:Stars SYSTEM "example.dtd">
```

Here the root element is defined.

```
    <!--Start of Data Proper -->
```

This line is a comment.

```
    <Glassblade:Stars xmlns:glassblade="http:www.
Glassblade.com">
    <Glassblade:Sol>Type 1</Glassblade:Sol>
    <Glassblade:Alpha>Type 2</Glassblade:Alpha>
</Glassblade:Stars>
```

Needless to say <Glassblade:Stars>, <Glassblade:Sol> and <Glassblade:Alpha> are all invented elements with no special relevance.

The Stylesheet

Defines the way elements are processed by translating one series of XML elements to another. This separation of data and the way it is formatted is fundamental to the way that XML operates - applying different stylesheets allows a "page" to be rendered in a way appropriate to the circumstances - whether it's a hi-resolution, high-bandwidth screen or a tiny mobile phone.

Stylesheets are written as a well-formed xml document and adhere to the same rules.

```
<?xml version="1.0"?>
<xsl:stylesheet
  xmlns:xsl="http://www.w3.org/TR/WD-xsl"
xmlns:fo="http://www.w3.org/TR/WD-xsl/FO"
xmlns:Glassblade="http://glassblade.com">
```

Define the XSL namespace along with any others needed- here we are referencing the "format object" namespace that allows access to a range of universal formatting tags.

```
<xsl:template match="/">
```

set the template pattern to the root element.

```
<fo:block font-size="18pt">
                <xsl:apply-template/>
```

Apply the formatting.

```
</fo:block>
</xsl:template>
<xsl:template match="Glassblade:Stars">
<fo:block font-size="12pt">
                <xsl:text>Star series</xsl:text>
<xsl:apply-template/>
```

Apply the formatting.

```
</fo:block>

</xsl:stylesheet>
```

The XSL is evolving into a powerful language in its own right with functions to map elements and define the behaviour of the XML documents. XML based applications will derive much of their functionality from this. With its rich flow control and interrogative elements such as "for-each", "if", "any" and "from" highly sophisticated patterns can be generated, allowing the raw XML document to be tailored to fit whichever environment is currently targeting it.

The Document Type Definition

The DTD provides rules for the various components of the XML document and how they interact. It also defines the grammar of the documents and the elements. The DTD has a sophisticated capacity to define attributes with substitution, enumeration, external references and a host of other operational directives.

The `<!ELEMENT elementname rule>` declaration is the basis for the structure of the DTD i.e.

```
<!ELEMENT StarDesc (#PCDATA)>
PCDATA represents Parsed Character data
```

The DTD also allows the creation of entities. These take the form of General entities which carry a role similar to programming constants like `<!ENTITY name "substituted characters">`.

For example `<!ENTITY Spectrum "&Colour">` could be invoked using

```
<Analysis>
&Spectrum; Red
</Analysis>
```

Parameter entities are analogous to variables in conventional programming with the form `<!ENTITIES % name "substituted characters >`

For example:

```
<!ENTITIES % sequence "(#PCDATA)">
<!ELEMENT NucleaseName %sequence>
```

Other types of entities include:

- external entities which allow external references such as a URI which allows the XML content referenced by the external entity (pointer?) to be copied into the current document;
- Unparsed entities declare non-xml content via a reference to an external data source with an NDATA (notational data) keyword that defines the type of unparsed data being referenced.

There are a range of datatypes available to the DTD including CDATA (characterdata), ID (id references) and NOTATION (extends the notation of the DTD).

And, as with XSL, there are DTD elements such as IGNORE that allow complex responses when used with parameter references etc.

Namespaces

This recent development allows prefixing an element with namespace: to ensure uniqueness. Declared using

`xmlns:`*`something`*

there can be several declared namespaces and scope is restricted to an elements opening and closing tags. The potential of namespaces is that it allows safe use of elements' names. For example if a physics department created an XML schema for visualizing particle interaction and had an element called <particle> it would not be unlikely that another Physicist would develop his own XML documents including his own <particle> element. Should they ever need to combine documents these elements would be said to collide. If however the namespace was declared with

```
<QMC:particle xmlns:QMC="HTTP://www.QMC.edu.uk>
```

then scoping the `<particle>` element as `<QMC:particle>` allows much better protection against collision.

Appendix D

MIME - Multi-purpose Internet Mail Extensions

Condensed from the Frequently Asked Questions page of the comp.mail.mime newsgroup.

MIME, the Multi-purpose Internet Mail Extensions, is a freely available set of specifications that offers a way to interchange text in languages with different character sets, and multi-media e-mail among many different computer systems that use Internet mail standards.

If you were bored with plain text e-mail messages, thanks to MIME you now can create and read e-mail messages containing these things:

- character sets other than US-ASCII
- enriched text
- images
- sounds
- other messages (reliably encapsulated)
- tar files
- PostScript
- pointers to FTPable files
- other stuff

MIME supports not only several pre-defined types of non-textual message contents, such as 8-bit 8000Hz-sampled mu-LAW audio, GIF image files, and PostScript programs, but also permits you to define your own types of message parts. Before MIME became widespread, you might have been able to create a message containing, say, a PostScript document and audio annotations, but more often then not, the message was encoded in a proprietary, non-transportable format. That meant that you couldn't easily handle the same message on another vendor's workstation, or even get it intact through a mail gateway in the first place. Now, depending on the completeness of your MIME-capable mail system, there's a good chance that it'll "just work".

One of the best things about MIME is that it's a "four-wheel drive protocol" (to borrow a description applied originally to PhoneNet by Einar Stefferud). MIME was carefully designed to survive many of the most bizarre variations of SMTP, UUCP, and other Procrustean mail transport protocols that like to slice, dice, and stretch the headers and bodies of e-mail messages.

As an example of how MIME could be used in the real world, Kim Dyer mails out her Simple Network Management Protocol-related newsletter, "The Al Stewart Chronicles" as multi-media e-mail messages in several forms:

- In a PostScript form, with beautiful typesetting and a two-column page layout, suitable for printing on a laser printer;
- In a "text/html" form (RFC 1866), suitable for examination via a WWW browser. (Formerly, text/richtext, another SGML-like markup language, was used.)
- In an ordinary, plain text, form.

MIME capability doesn't automatically confer interoperability with the rest of the world. Any random data can be mapped into MIME one way or another, but some consideration needs to be given to the target audiences.

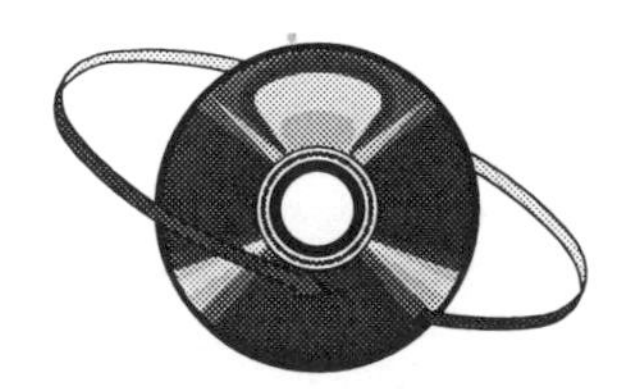

Appendix E

Glossary and WAP-related Acronyms

3G
See UMTS

Broadband phone
A third-generation (3G) mobile phone that has higher access speeds than currently available.

Card
A card is a single block of WML code which can contain basic text or navigation items. Several cards are known as a 'deck'.

Client
A device (or application) that initiates a request for connection with a server.

Deck
A set of WML cards which is first loaded whenever the user or WAP device requests a URL.

Device
A network entity that is capable of sending and receiving packets of information and has a unique device address.

DTD
Acronym for Document Type Definition. A DTD definition defines the names and contents of elements within a document such as an XML file. Also dictates which elements can be nested within others and defines other rules. Can be viewed as a way of defining your own mark-up tags.

Dual Band
Most mobile phone networks throughout the world transmit on the radio frequency band of 900MHz, including BT cellnet and Vodafone. However, a few, such as Orange and One2One in the UK, transmit on 1800MHz. Phones which can work on both frequencies are known as Dual-band handsets.

Element
An element specifies the markup and structural information inside a WML deck. Elements which have start and end tags such as the <p> and </p> (paragraph) tags are known as containers. If, however, an element exists on it's own, the forwardslash must follow the tag e.g. the
 (line break) tag.

GSM
An acronym for General Systeme Mobile. Nearly all mobile-networks currently operate on GSM, but they may use different bands. Will be replaced by UMTS.

GPRS
GPRS is an acronym for General Packet Radio Service. It is a packet-based wireless communication service that allows data exchange rates of up to 114 Kbps, and continuous connection to the Internet known as an "always-on mode". GPRS is at the basis of third-generation.

HDML
The acronym for Handheld Devices Markup Language which is now called the Wireless Markup Language (WML)

HTML
Acronym for HyperText Mark-up Language. A tag-based language that defines the appearance of elements within a document. Most commonly used in the creation of World Wide Web pages.

HTTP
The Hypertext Transfer Protocol is the set of rules for exchanging files on the World Wide Web. It is an application-based protocol.

JavaScript/JScript

A scripting language used in web browsers as a client-side technology and as a server-side integration tool.

m-commerce

Short for mobile-commerce in the same way that e-commerce is used for electronic commerce. M-commerce is a subset of e-commerce. It generally refers to the ability to conduct financial transactions over a mobile device, such as a WAP phone or PDA.

Menu

A term now used to describe the software options on a mobile phone. Came about as a result of its widespread usage in relation to PCs.

MIP

An acronym for Mobile Internet Provider. MIPS are analogous to and, indeed, may be the same as ISPs (Internet Service Providers).

Network

The company that actually transmits your phone calls and messages between handsets, and who own the transmitters.

PDA

An acronym for Personal Digital Assistant. A hand-held, and hence mobile, device such as the Psion or Palm Pilot.

Phonebook

Your personal list of phone numbers and contact details saved on the SIM Card or in the handset's memory.

Predictive text

Software on the handset which tries to guess which of the letters on a button you intend to select when writing an SMS or Email (three or more letters are usually assigned to one button). The same software may try and guess whole words based on the rest of the sentence. Also known as Easy text and T9 after the commercial name of the software.

SDK
An acronym for System Development Kit. They are used for the development of wireless applications using languages such as WML. Kist are available from many sources, the most common being Nokia and Ericsson.

Server
A device (or application) that passively waits for connection requests from one or more clients. A server may accept or reject a connection request from a client.

Simulator
A piece of software used to emulate what you would see on another physical device. The most common usage is to enable you to see a WAP site with your desktop PC.

SGML
An acronym for Standardised Generalised Markup Language, the standard out of which HTML and WML were derived.

SIM Card
An electronic chip which stores your personal account details and phone number. You can swap these between phones provided they aren't SIM locked and the two phones are not single-band phones on networks working on a different frequency.

SMS
An acronym for Short Message Service. A messaging service supported by many mobile phones that allows short text messages to be sent between mobile devices. It is not an interactive protocol like WAP.

Tri-Band
A phone which works on the two dual-band frequencies of 900MHz and 1800MHz, and also on the American frequency of 1900MHz.

User Agent
Another name for a WAP device, microbrowser or web browser,

that interprets content coded in formats such as WML, WMLScript, and HTML. This may include textual browsers, voice browsers, search engines, etc.

UMTS
Short for Universal Mobile Telephone System, which is the name of a new mobile networking standard that will replace GSM in a couple of years. UMTS has data speeds many hundreds of times faster than GSM and is commonly known as Third generation or 3G.

URL
An acronym for Uniform Resource Locator. URLs are addresses of web-based resources. It can refer to static pages and to applications (scripts).

VBScript
VBScript is a Microsoft scripting (development) language that is based on Visual Basic. VBScript can be used in the client and on the server.

W3C
Acronym for the World Wide Web Consortium, the Web's main standard setting body.

WAE
An acronym for Wireless Application Environment. It specifies a general-purpose application environment based fundamentally on Web technologies and is part of the WAP standard.

WAP
WAP is the Wireless Application Protocol. An agreed specification for a set of communication protocols which standardise the way a wireless mobile devices can be used for Internet access. WAP was originally conceived by Ericsson, Motorola, Nokia and Unwired Planet (who are now called Phone.com).

WAP Device
A device which renders content written in WML.

WAP Gateway
A WAP gateway is an interface with the WAP device on one side, and the web server on the other. The gateway's job is to convert the web data into a format which can be displayed on the WAP device. On the server side, it can provide the web server with information about the WAP device.

WAP Server
A WAP server is the equivalent to a web server. Most WAP servers are HTTP servers. Some WAP servers can also acts as gateways, allowing for the serving of web content and WML.

WASP
Acronym for Wireless Applications Service Provider - a company which provides content and applications for wireless devices.

WML
Wireless Markup Language is a tag-based language used for delivering data to WAP devices, and is HTML-like in its appearance and use.

WMLScript
A mobile scripting language similar to JavaScript in terms of syntax

WSP
An acronym for Wireless Session Protocol. It provides the upper-level application layer of WAP with a consistent interface for two session services.

WTAI
An acronym for Wireless Telephony Application Interface. The WTAI specification describes standard telephony-specific extensions to WAE.

XML
An acronym for Extensible Markup Language. The W3C's standard for Internet Mark-up Languages. WML is one of these languages, and is a 'dialect' of SGML. XML describes the

structure of content, unlike HTML that describes how pages are "marked-up".

API	Application Programming Interface
CC/PP	Composite Capability/ Preferences Profiles
CGI	Common Gateway Interface
CPI	Capability and Preference Information
DCS	Digital Communications System
DTMF	Dual Tone Multi-Frequency
GSM	Global System for Mobile Communication
HTML	Hyper-Text Markup Language
HTTP	Hyper-Text Transfer Protocol
P3P	Platform for Privacy Preferences Project
PCS	Personal Communications System
PDA	Personal Digital Assistant
PDC	Personal Digital Cellular
PICS	Protocol Implementation Conformance Statement
RDF	Resource Description Framework
SiRPAC	Simple RDF Parser and Compiler
SSL	Secure Sockets Layer
TLS	Transport Layer Security
UAProf	User Agent Profile
URI	Uniform Resource Identifier
URL	Uniform Resource Locator
W3C	World Wide Web Consortium
WAE	Wireless Application Environment
WAP	Wireless Application Protocol
WBXML	WAP Binary XML
WCMP	Wireless Control Message Protocol
WDP	Wireless Datagram Protocol
WML	Wireless Markup Language
WSP	Wireless Session Protocol
WTA	Wireless Telephony Application
WTAI	Wireless Telephony Applications Interface
WTLS	Wireless Transport Layer Security
WTP	Wireless Transaction Protocol
XHTML	Extensible HyperText Markup Language
XML	Extensible Markup Language

The Net-Works Guide to
Marketing Your Website

Simply creating a website and putting it on the Internet is not enough to generate online sales, no matter how good it looks. In Cyberspace there simply isn't any passing traffic. The customer is king, and they will selectively choose which sites they are going to visit. Indeed, without a well thought-out and successfully implemented marketing plan, the majority of your potential customers will not even know that you are out there.

What is more, even if you have a few years of 'real life' marketing experience under your belt, you have a lot to learn, and unlearn, before attempting to market your website. The Internet is a very different medium from those you may be used to, and your marketing strategy will have to adapt accordingly if you want to enjoy any kind of success.

Online Marketing Made Easy

The Net-Works Guide to Marketing Your Website will show you how to construct and deliver a successful promotional strategy. It covers everything from the basics of linking to other sites and search engine registration, through to referral sales and associate programs. No stone is left unturned in the quest for more visitors. You will learn:

- How to promote your site in newsgroups and chat rooms without being flamed,
- The importance of META tags and how to use them,
- How to build customer loyalty,
- What is meant by 'sticky content' and how to write it,
- Why e-zines, newsletters and bait pages are important,
- Ways of promoting your site in the traditional media,

Anyone reading this book, and putting its easy-to-follow, non-technical advice into operation, can expect a rapid increase in the number of hits to their website.

Tim Ireland 112 pages £7.95

Starting and Running a Business on the Internet

Do you want to:

- ✔Sell your goods all over the world without leaving your office chair?
- ✔Tap the fastest growing and most affluent market ever?
- ✔Slash your marketing and advertising costs?
- ✔Talk to the other side of the world for free?
- ✔Have access to strategic information only the biggest companies could afford?

Then your business should be on the Internet!

Companies are already cutting costs, improving customer support and reaching hitherto untapped markets via the Internet. They have realised the potential for this exciting new commercial arena and they've grabbed the opportunity with both hands. Now you can join in the fun of what is still a 'ground floor' opportunity.

Starting and Running a Business on The Internet offers realistic and practical advice for any existing business or budding 'Cyberpreneur'. It also:

- ❑ Helps you get started QUICKLY and CHEAPLY.
- ❑ Tells you which sites 'work', which don't and, more importantly, WHY!
- ❑ Details how to PROMOTE your business online.
- ❑ Shows you how to stay ahead of your competitors.
- ❑ Warns you of the major PITFALLS and shows you how to AVOID them.
- ❑ Highlights important issues like CREDIT CARD handling and site SECURITY.

Alex Kiam 112 pages £6.95

The Net-Works Guide to
Creating a Website

The World Wide Web has established itself as an important business and communications tool. With hundreds of millions of computer users around the globe now relying on the Web as their primary source of information, entertainment and shopping, it cannot be ignored.

Whether it is to showcase your business and its products, or to present information on your favourite hobby or sport, creating your own Web site is an exciting development. But unless you're familiar with graphics programs and HTML (the "native language" of the Web), as well as how to upload files onto the Internet, creating your Web pages can also be very frustrating! But it doesn't have to be that way.

Web Publishing Made Easy

This book, written by a Website design and marketing consultant, will help demystify the process of creating and publishing a Web site. In it you will learn:

- How to research and plan your site,
- What free tools are available that make producing your own Web site child's play (and where to find them),
- How to create your own dazzling graphics, using a variety of cheap or free computer graphics programs,
- How to put it all together to achieve a great look with a minimum of fuss, and
- How to promote your Web site and attract other Internet users to it.

Advanced Web Design Issues

In addition to canvassing the basics of creating your first Web site, the author also discusses more advanced Web design issues... how to focus your Web site content for your target audience... how to minimise the time taken for your Web site to download... and what lies ahead for the Web and eCommerce... etc., etc.

Tim Ireland 112 pages £6.95

The Internet for Managers

Embrace the Internet,
or Lose Your Job?

Every modern-thinking company executive recognises the words of Internet gurus when they say, "In five years time there will be two types of company; those on the Internet and those who have ceased trading".

But despite this, many managers still view the Internet as a scary place. It appears to be the preserve of the technical wizard and even the acronyms seem incomprehensible.

The Internet for Managers tackles this problem head on and gives all levels of management enough information to 'hold their own' in meetings and to take on board the issues which affect their spheres of responsibility. It explains all the key concepts from the basics of Web browsing through to the establishment of corporate security policies in clear, non-technical terms.

Learn about:

Joining the Net... Constructing Websites... Electronic Trading... Marketing Online... Intranets, Teleworking and the Virtual Corporation... Technical Issues... Security Problems... Language Difficulties... Legal Implications...The Government's Attitude... Partnerships and Strategic Alliances... and much more.

Impress Your Colleagues

This concise and enjoyable executive summary will leave you highly informed on the challenges posed by the Internet and e-commerce. Read this book and your less-knowledgeable colleagues and competitors will be left way behind, scrambling to stay competitive.

RRC Penfold 192 pages £12.95

Book Ordering

To order any of these books, please order from our secure website at **www.net-works.co.uk** or complete the form below (or use a plain piece of paper) and send to:

Europe/Asia

TTL, PO Box 200, Harrogate HG1 2YR, England (or fax to 01423-526035, or email: sales@net-works.co.uk).

USA/Canada

Trafalgar Square, PO Box 257, Howe Hill Road, North Pomfret, Vermont 05053 (or fax to 802-457-1913, call toll free 800-423-4525, or email: tsquare@sover.net)

Postage and handling charge:

UK - £1 for first book, and 50p for each additional book

USA - $5 for first book, and $2 for each additional book (all shipments by UPS, please provide street address).

Elsewhere - £3 for first book, and £1.50 for each additional book via surface post (for airmail and courier rates, please fax or email for a price quote)

Book	Qty	Price
	Postage	
	Total:	

☐ I enclose payment for £______

☐ Please debit my VISA/AMEX/MASTERCARD

Number: ☐☐☐☐ ☐☐☐☐ ☐☐☐☐ ☐☐☐☐

Expiry Date: ☐☐☐☐ Signature: Date:

Name: ______________________

Address: ______________________

____

____

Postcode/Zip: ______________________

Telephone/Email: ______________________

wap+wml